CHEMICAL KINETICS
AND REACTION MECHANISMS

McGraw-Hill Series in Advanced Chemistry

Ballhausen: *Introduction to Ligand Field Theory*
Benson: *The Foundations of Chemical Kinetics*
Davidson: *Statistical Mechanics*
Dean: *Flame Photometry*
Dewar: *The Molecular Orbital Theory of Organic Chemistry*
Eliel: *Stereochemistry of Carbon Compounds*
Espenson: *Chemical Kinetics and Reaction Mechanisms*
Fitts: *Nonequilibrium Thermodynamics*
Fitts: *Vector Analysis in Chemistry*
Hammett: *Physical Organic Chemistry*
Helfferich: *Ion Exchange*
Hine: *Physical Organic Chemistry*
Jencks: *Catalysis in Chemistry and Enzymology*
Kan: *Organic Photochemistry*
Laidler: *Theories of Chemical Reaction Rates*
Laitinen and Harris: *Chemical Analysis*
McDowell: *Mass Spectrometry*
Mandelkern: *Crystallization of Polymers*
March: *Advanced Organic Chemistry: Reactions, Mechanisms, and Structure*
Memory: *Quantum Theory of Magnetic Resonance Parameters*
Pitzer and Brewer: (Revision of *Lewis and Randall*) *Thermodynamics*
Plowman: *Enzyme Kinetics*
Pople and Beveridge: *Approximate Molecular Orbital Theory*
Pople, Schneider, and Bernstein: *High-Resolution Nuclear Magnetic Resonance*
Pryor: *Free Radicals*
Raaen, Ropp, and Raaen: *Carbon 14*
Shriver: *The Manipulation of Air-Sensitive Compounds*
Siggia: *Survey of Analytical Chemistry*
Wertz and Bolton: *Electron Spin Resonance: Elementary Theory and Practical Applications*
Wiberg: *Laboratory Technique in Organic Chemistry*

CHEMICAL KINETICS AND REACTION MECHANISMS

James H. Espenson

Professor of Chemistry
Iowa State University

McGraw-Hill Book Company

New York   St. Louis   San Francisco   Auckland   Bogotá   Hamburg
Johannesburg   London   Madrid   Mexico   Montreal   New Delhi
Panama   Paris   São Paulo   Singapore   Sydney   Tokyo   Toronto

This book was set in Times Roman by Automated Composition Service, Inc.
The editors were Jay Ricci and James W. Bradley;
the production supervisor was Diane Renda.
The drawings were done by Automated Composition Service, Inc.
Fairfield Graphics was printer and binder.

CHEMICAL KINETICS AND REACTION MECHANISMS

234567890 FGFG 8987654321

Library of Congress Cataloging in Publication Data

Espenson, James H.
    Chemical kinetics and reaction mechanisms.

    (McGraw-Hill series in advanced chemistry)
    Includes index.
    1. Chemical reaction, Rate of. 2. Chemical
reaction, Conditions and laws of. I. Title.
QD502.E86   541.3'94   80-17206
ISBN 0-07-019667-2

CONTENTS

Preface ix

Chapter 1 Reactions and Reaction Rates 1

1-1 Net Reactions and Reaction Rates 2
1-2 Elementary Reactions and Reaction Mechanisms 4
1-3 Order of a Reaction 5
1-4 Factors Influencing Reaction Rates 7
1-5 Practical Kinetics 8

Chapter 2 Reactions with a Simple Kinetic Form 12

2-1 First-Order and Pseudo-First-Order Reactions 12
2-2 Second-Order Kinetics 16
2-3 Equivalent Concentrations in Second-Order Kinetics 20
2-4 Use of Physical Properties in First- and Second-Order Kinetics 22
2-5 Methods When the Infinity Reading or "End Point" Is Unknown 24
2-6 Other Simple Kinetic Forms 30
2-7 Rate Expressions and the Method of Flooding 30
2-8 The Experimental Determination of Reaction Orders 32
2-9 Multiterm Rate Expressions 35

Chapter 3 Kinetics of Complex Reactions—Reversible and Concurrent Reactions 42

3-1 Reversible First-Order Reactions 42
3-2 Opposing Reactions of Higher Order 45
3-3 "Concentration-Jump" Methods for Reversible Equilibria 48
3-4 Exchange Reactions 50
3-5 Parallel and Concurrent First-Order Reactions 55
3-6 Concurrent Reactions of Mixtures 57
3-7 Competition Experiments for Reaction Intermediates 58

Chapter 4 Consecutive Reactions and Reaction Intermediates   65

4-1   Consecutive First-Order Reactions   65
4-2   Dual Solutions in Consecutive Reactions   69
4-3   A Consecutive Reaction with a Reversible Step   71
4-4   The Steady-State Approximation   72
4-5   Limiting Forms; the Rate-Limiting Step   76
4-6   Direct versus Sequential Reaction   78
4-7   An Intermediate Which Reacts in Steps Having Different Reaction Orders   79
4-8   Kinetic Equations for Catalyzed and Enzyme-Catalyzed Reactions   80
4-9   Numerical Solutions to Rate Equations   83

Chapter 5 Deduction of Reaction Mechanisms   89

5-1   The Activated Complex or Transition State   89
5-2   Mechanistic Interpretation of Rate Laws   90
5-3   Equivalent Kinetic Expressions   94
5-4   Parallel Pathways   97
5-5   Successive Steps   98
5-6   Preequilibria   101

Chapter 6 Reaction Energetics and Chemical Kinetics   116

6-1   The Variation of Rate Constant with Temperature   116
6-2   Activation Parameters   118
6-3   The Temperature Dependence of Composite Rate Constants   121
6-4   Relation of Forward and Reverse Reaction Rates   123
6-5   Connections between Kinetics and Thermodynamics   124
6-6   The Heat Capacity of Activation   125
6-7   The Principle of Microscopic Reversibility   128

Chapter 7 Chain Reactions   134

7-1   Characteristics of Chain Reactions   134
7-2   The Decomposition of Acetaldehyde   135
7-3   The Hydrogen-Bromine Reaction   137
7-4   The Photochemical Hydrogen-Bromine Reaction   139
7-5   Free-Radical Halogenations of Hydrocarbons   140
7-6   Steady-State Approximation in Chain Reactions   142
7-7   Branching Chain Reactions   142
7-8   Oscillating Reactions   143

Chapter 8 Theories of Elementary Reaction Rates   150

8-1   Collision Theory for Bimolecular Reactions   150
8-2   Steric Factors and Energy Terms in Collision Theory   152
8-3   Activated Complex Theory   153
8-4   Derivation of the Activated Complex Theory   153

8-5 Application of Activated Complex Theory to Bimolecular
   Reactions   155
8-6 Calculations Based on Activated Complex Theory   157
8-7 Kinetics of Unimolecular Gas-Phase Reactions   159

Chapter 9 Reactions in Solution   166

9-1 The Nature of Reactions in a Solvent   166
9-2 The Rates of Diffusion-Controlled Reactions   168
9-3 Applications of Activated Complex Theory   170
9-4 Solvent Effects on Polar and Ionic Reactions   170
9-5 Salt Effects on Second-Order Ionic Reactions   172
9-6 Salt Effects on Other Reaction Orders   175
9-7 Salt Effects and Reaction Mechanisms   176
9-8 Influence of Pressure on Solution Reactions   177
9-9 Concentration Units in Solution Reactions   179

Chapter 10 Reactions at Extreme Rates   182

10-1 Survey of Methods for Very Fast Reactions   182
10-2 Flow Methods for Rapid Reactions   183
10-3 Relaxation Methods   184
10-4 Kinetic Equations for Relaxation Kinetics   185
10-5 Magnetic Resonance Methods   187
10-6 Flash Photolysis   188
10-7 Pulse Radiolysis   189

Chapter 11 Extrakinetic Probes of Mechanism   193

11-1 Linear Free-Energy Correlations   193
11-2 The Hammett Relation   195
11-3 The Marcus Relation for Electron Transfer   197
11-4 Acid-Base Catalysis   198
11-5 Stereochemistry and the Activation Process   202
11-6 Isotope Rate Effects   203
   Appendix: A Derivation of the Marcus Equation for
   Electron Transfer   204

Answers to Selected Numerical Problems   209
Index   213

PREFACE

Chemical kinetics and its supporting techniques have proved to be valuable approaches to the study of reaction mechanisms and to an understanding of chemical reactivity. This brief text is intended to provide an introduction to many aspects of the broad field of chemical kinetics with an emphasis upon chemical and mechanistic interpretations. To realize that goal, specific examples from the literature have been used as illustrations whenever possible, rather than more abstract formulations involving the ubiquitous and unspecified substances A, B, etc.

The chemical examples have been chosen with two points in mind: first, that they illustrate the subject under discussion in a straightforward manner; second, that the chemistry itself (the main point of scientific interest, but perhaps incidental to the considerations at hand) be not so unfamiliar to most readers as to complicate the issue.

Organized accounts of the mechanisms of organic or inorganic or organometallic or biochemical reactions are not included in this text. Those subjects are properly left to courses in the fields. The goal here is to show how mechanistic results can be obtained and evaluated, rather than to consider the results for their own sake.

This text is based upon a one-term course at Iowa State University which enrolls graduate students from all disciplines of chemistry as well as the occasional undergraduate and nonchemist. It should also serve as an introduction to chemical kinetics for research workers in other fields and as a supplementary text in kinetics for courses in organic, inorganic, or biochemical mechanisms. Although many readers are likely to have had an undergraduate course in physical chemistry, extensive prior exposure to kinetics is not assumed. Indeed, this text could be used as one of the texts in such a course.

Problems accompany each chapter. They are to be regarded as an integral part of the course of study, and those who wish the full benefit from this text will give them due effort. A number of aspects which are not developed explicitly in the text are elicited in the problem solutions.

In keeping with the theme of this text, the treatment of the theory of chemical kinetics is limited to that needed to provide the experimental chemist with a summary of selected aspects of immediate interest. Applications to polymers, engineering reactor design, the dynamics of elementary reactions, and scattering theory are omitted; and the kinetics of heterogeneous and enzymatic reactions are considered only briefly.

It is a pleasure to acknowledge the assistance of many individuals, including Edward L. King, who first introduced me to the subject; Gilbert Gordon, who generously shared with me an extensive set of notes he prepared for a similar course; authors of numerous research articles and books; my research students as well as those enrolled in Chemistry 528; and (with apologies) unnamed individuals whose oral or written contributions I may inadvertently have failed to acknowledge. I am grateful for numerous suggestions concerning the manuscript made by Drs. J. P. Birk, W. R. Bushey, E. S. Gould, E. L. King, H. Ogino, and R. C. Thompson.

Complexity in a chemical reaction may aid, rather than hinder, in the determination of a mechanism. To quote one author:*

It is a mistake to confound strangeness with mystery. The most commonplace crime is often the most mysterious, because it presents no new or special features from which deductions may be drawn. This murder would have been infinitely more difficult to unravel had the body of the victim been simply found lying in the roadway without any of those *outré* and sensational accompaniments which have rendered it so remarkable. These strange details, far from making the case more difficult, have really had the effect of making it less so.

James H. Espenson

*A. C. Doyle, "A Study in Scarlet," in "The Complete Sherlock Holmes," vol. 1, Doubleday, New York, 1953, p. 44, as quoted by J. F. Bunnett in E. S. Lewis (ed.), "Investigations of Rates and Mechanisms of Reactions," vol. VI, part 1 of the series "Techniques of Chemistry," Wiley-Interscience, New York, 1974, p. 480.

CHEMICAL KINETICS
AND REACTION MECHANISMS

REACTIONS AND REACTION RATES

Chemical kinetics is the study of the rates of chemical reactions. Such studies are closely related to the development of an understanding of reaction mechanisms. Indeed, chemical kinetics and reaction mechanisms are often regarded as synonymous. We do stop short, however, of stating that a kinetic study, even a very thorough one, constitutes a determination of the reaction mechanism. The latter is but a scientific postulate open to revision in the light of new data and newer theories of reactivity.

It will often be the case that a proposed mechanism is not unique, and other postulates may account for the data equally well. Research workers concerned with reaction mechanisms must develop the habit of spotting such alternatives with some facility. They must also be willing to take a critical attitude toward mechanistic proposals, especially their own, in an effort to narrow the range of possibilities.

A mechanism is never really "proved," but sufficient and sufficiently definitive experiments may eliminate some alternatives as being impossible or unlikely, perhaps leaving but one credible mechanism remaining. Lewis[1][†] has pointed out that the subject of reaction mechanism deserves no special censure on this point; for no model or hypothesis in science which has been advanced to explain a set of observations is ever proved. The best one does is exactly what one does in a mechanistic study: Consider the proposal from every possible view and try to design tests of a "critical" nature, tests that will probe every assumption and every assertion.

The term "reaction mechanism" takes on different shades of meaning in different contexts. From its earliest development, "mechanism" has been taken to include the molecular events which occur during a reaction. In that context one would seek to learn whether a given chemical reaction occurs in a single molecular process (a so-

[†]Superscript numbers are those of references listed at the ends of chapters.

called elementary reaction) or in several processes. If the latter, do the molecular steps occur as concurrent alternatives or in succession along a single pathway? What is the stereochemical arrangement of the *activated complex*, the state of maximum potential energy through which the system passes in its key phase, the *rate-limiting step*?

The list of similar questions could be extended considerably, but it is not necessary to elaborate at this point; the questions are familiar to persons with training in elementary chemistry. Chemical kinetics and its subsidiary tools provide the means by which answers of varying degree of certainty may be formulated.

In some cases the resolution of mechanistic questions at this level is thought to be reasonably complete. That may be so because the nature of a particular chemical system makes the mechanism especially amenable to resolution and because considerable effort has been devoted to its study. On the other hand, there are many processes, including some which have been studied quite extensively, for which a satisfactory resolution of the alternatives has not been effected.

Aspects of mechanism such as those just mentioned are most often in the minds of persons interested in understanding reactions of substances of interest to them. Typical organic and inorganic chemists and biochemists ask such questions about their reactions.

On a different level, however, the term "reaction mechanism" has also come to mean the delineation of the fine molecular details of a reaction—an understanding of the dynamics of each collision step, the energy transfer processes, the excited states involved, and the exact potential-energy surfaces for each elementary reaction. Those aspects are often referred to nowadays as *chemical dynamics*, and they are most often of concern to physical chemists. We shall touch on some of them, but they are not treated in detail in this text.

It is useful to start our account of chemical reactions, reaction rates, and reaction mechanisms by summarizing in the first chapter a number of items familiar to most readers from their earlier study of chemistry.

1-1 NET REACTIONS AND REACTION RATES

The equation for a net chemical reaction represents the overall stoichiometric transformation of reactants into products without addressing questions related to reaction mechanism. Thus iron(II) is oxidized by thallium(III) according to the equation[†]

$$2Fe^{2+} + Tl^{3+} = 2Fe^{3+} + Tl^{+} \tag{1-1}$$

Mechanistic conclusions are not to be drawn from Eq. (1-1), and it is certainly not appropriate to infer, in the absence of direct kinetic determinations, that the reaction rate will be proportional to the product $[Tl^{3+}][Fe^{2+}]^2$. It is true that the rate

---

[†]A useful convention is the use of the equality in a net reaction. Single and double arrows are respectively reserved for irreversible and reversible elementary reactions in a chemical reaction mechanism.

of reaction will almost certainly depend upon the concentrations of the reactants, but not necessarily in that manner.

The rate of a reaction in a constant-volume batch reactor is expressed as the change in concentration of any reactant or product per unit time. Such a reactor constitutes the common design usually preferred for kinetic studies directed toward the investigation of chemical reaction mechanisms, and the treatment given here is limited to that case. For expression of the rate of reaction, the second is the unit of time used by nearly all workers in different areas of chemical kinetics. Most journals now insist that it be used, and that has produced a highly desirable degree of standardization in the literature. Many workers use molarity (mol dm^{-3} or mol L^{-1}) as the preferred concentration unit, both for expression of the rate of reaction in units mol dm^{-3} s^{-1} and for the concentrations of the substances upon which the rate depends. The use of the molar concentration unit is widespread in solution chemistry. For gas-phase reactions, the units mol cm^{-3} and molecules cm^{-3} for concentrations are sometimes found, as are pressure units such as bar, atmosphere, torr, and pascal.

In the example cited previously, the rate of reaction between thallium(III) and iron(II) is, indeed, *not* proportional to the product $[Tl^{3+}][Fe^{2+}]^2$. The experimental kinetic data show that the reaction rate, which can be expressed as $-d[Tl^{3+}]/dt$, is given by

$$\frac{-d[Tl^{3+}]}{dt} = k[Tl^{3+}][Fe^{2+}] \tag{1-2}$$

Since thallium(III) is being consumed in the reaction, its concentration derivative is negative. The explicit negative sign in Eq. (1-2) makes the reaction rate itself positive. As a consequence, the proportionality constant or rate constant k has a positive value. An equation, such as (1-2), that gives the reaction rate as a function of concentration, usually at constant temperature, is referred to as a *rate law*. The determination of its form is one of the initial goals of a kinetic study. Contained within the rate law is certain information about the mechanism, as will be considered in due course.

Any of the reactants or products could equally well have been chosen as the basis for expression of the reaction rate. Any derivative which is convenient for the purpose at hand may be chosen. In general, however, the following is not a very satisfactory statement: Rate = $k[Tl^{3+}][Fe^{2+}]$. In this form a numerical ambiguity arises because the concentration variable upon which "Rate" has been based has not been explicitly stated. Various possible definitions are interrelated by the stoichiometry of the net reaction, as in Eq. (1-3).

$$\frac{-1}{2}\frac{d[Fe^{2+}]}{dt} = \frac{-d[Tl^{3+}]}{dt} = \frac{+1}{2}\frac{d[Fe^{3+}]}{dt} = \frac{+d[Tl^{+}]}{dt} \tag{1-3}$$

The expressions in Eq. (1-3) follow from the overall reaction and its stoichiometric coefficients. The careful worker will explicitly define the usage chosen for the case under consideration to avoid uncertainty in the quantity being considered. The validity of Eq. (1-3) also depends upon a further condition: that no intermediates build up to a "significant" concentration. That will be the case if the following mass balance relations are satisfied:

$$\tfrac{1}{2}([Fe^{2+}]_0 - [Fe^{2+}]_t) = ([Tl^{3+}]_0 - [Tl^{3+}]_t)$$

$$= \tfrac{1}{2}([Fe^{3+}]_t - [Fe^{3+}]_0) = ([Tl^+]_t - [Tl^+]_0) \qquad (1\text{-}4)$$

That is not to say that an intermediate such as Tl^{2+} or Fe^{4+} is not involved, but its concentration remains low relative to reactants and products. Indeed, since the stoichiometry of the net reaction and that of the rate law are not the same, an intermediate is necessarily involved.

Few reactions occur in a single step, so the question of reaction intermediates will be one of continuing interest to us. The ability of the investigator to detect a difference, for example, between $[Tl^{3+}]_0 - [Tl^{3+}]_t$ and $[Tl^+]_t - [Tl^+]_0$—such a discrepancy represents the concentration of Tl^{2+} at any time, however minute—depends upon the accuracy of the two analyses; a difference thought to be insignificant may become measurable with the development of a better analytical procedure. Also, a specific analysis for the intermediate may be developed and permit detection at levels far below those significant on the basis of differences.

1-2 ELEMENTARY REACTIONS AND REACTION MECHANISMS

We have seen that the term "reaction mechanism" is used in several contexts. For our immediate purpose we take it to represent the collection of individual molecular steps or elementary reactions comprising the net reaction.

The molecularity of an elementary reaction is the number of solute species which come together to form the activated complex. The rate of an elementary reaction is simply that given by the stoichiometry of the reaction itself. Elementary reactions are generally unimolecular or bimolecular, or perhaps occasionally termolecular.[†] If the dimerization of methyl radicals is taken to be an elementary reaction, as is believed to be the case, then its rate will be directly proportional to $[CH_3]^2$. In writing the form of the rate law for an elementary reaction, one must again be careful to specify the definition of rate in order to avoid the numerical ambiguity mentioned previously. A suitable notation for doing so has been devised. If this reaction is written as shown in Eq. (1-5),

$$2CH_3 \xrightarrow{k_1} C_2H_6 \qquad (1\text{-}5)$$

then either a specific differential equation such as Eq. (1-6) should accompany the re-

$$\frac{-d[CH_3]}{dt} = 2k_1[CH_3]^2 \qquad (1\text{-}6)$$

action or a convention should be adopted to obviate the need for defining equations. One convention widely but not uniformly adopted is the following: The elementary

---

[†]For present purposes we also include groupings of elementary reactions which are not independently resolved by the kinetics. Consequently, the defining expressions in Eqs. (1-7) and (1-8) are not limited to genuine unimolecular and bimolecular forms.

reaction of Eq. (1-7) has associated with it the rate law in (Eq. 1-8).

$$aA + bB + \cdots \xrightarrow{k} cC + dD + \cdots \tag{1-7}$$

$$\frac{-1}{a}\frac{d[A]}{dt} = \frac{-1}{b}\frac{d[B]}{dt} = \frac{+1}{c}\frac{d[C]}{dt} = \frac{+1}{d}\frac{d[D]}{dt} = k[A]^a[B]^b \tag{1-8}$$

1-3 ORDER OF A REACTION

The order of a reaction is the sum of the exponents of concentration factors in an experimental rate law. One can also refer to the reaction order with respect to a particular species. Reaction order is thus an experimental result for a chemical reaction under specified conditions. It is distinct from molecularity, which refers to the number of solute species reacting in an *elementary* process. Consider the reaction

$$5Br^- + BrO_3^- + 6H^+ = 3Br_2 + 3H_2O \tag{1-9}$$

which has a rate given by

$$\frac{-d[BrO_3^-]}{dt} = k[BrO_3^-][Br^-][H^+]^2 \tag{1-10}$$

The reaction is said to be first-order with respect to $[BrO_3^-]$ and $[Br^-]$ and second-order with respect to $[H^+]$. Overall, the reaction is fourth-order.

A mathematical definition of the reaction order with respect to the concentration of one particular substance i, C_i, is

$$\text{Order with respect to reagent } i = \left(\frac{\partial \log \text{rate}}{\partial \log C_i}\right)_{C_j} \tag{1-11}$$

where the reaction rate is evaluated under conditions such that the concentrations of the other substances, C_j, are constant. In the particular example cited, the order with respect to bromate ion concentration, for example, is $\partial \log \text{rate}/\partial \log [BrO_3^-] = 1.00$.

Were there a means of measuring the rates of reaction in a series of runs having constant $[H^+]$ and constant $[Br^-]$, a plot of log rate against log $[BrO_3^-]$ should define a straight line of slope 1.00. When the situation is not quite that simple, the definition of Eq. (1-11) and the graph of log rate versus log concentration suggested by the equation can often prove helpful.

Consider a reaction for which the reaction orders are not simple whole numbers. The following ligand substitution reaction is an example because its reaction rate contains two terms. Although each term individually has clearly defined orders of reaction, the reaction rate itself does not.

$$CrCl^{2+} = Cr^{3+} + Cl^- \tag{1-12}$$

$$\frac{-d[CrCl^{2+}]}{dt} = \left(k_1 + \frac{k_2}{[H^+]}\right)[CrCl^{2+}] \tag{1-13}$$

This reaction is first-order with respect to $[CrCl^{2+}]$, but clearly the order with respect to $[H^+]$ does not have a simple value. Rather, the *apparent order* with respect to $[H^+]$ varies with $[H^+]$. Thus at low $[H^+]$, such that $k_2/[H^+] \gg k_1$, the second term in the rate law dominates and $-d[CrCl^{2+}]/dt \cong k_2[CrCl^{2+}]/[H^+]$. The reaction order with respect to $[H^+]$ approaches -1 (inverse first-order) in this limit. On the other hand, at high $[H^+]$, such that the reverse inequality holds, the rate is approximated as $-d[CrCl^{2+}]/dt \cong k_1[CrCl^{2+}]$. The reaction has in this limit a zero-order dependence on $[H^+]$.

Reaction (1-12) is thus one in which one of the reaction orders, that with respect to $[H^+]$, varies with the concentration of H^+. From the preceding discussion a plot of log (rate at constant $[CrCl^{2+}]$) versus log $[H^+]$ should be a curve which approaches slopes of -1 and 0 at the respective low and high limits of $[H^+]$.

One means of presenting the experimental data so as to obtain the desired ordinate of log rate is to rearrange Eq. (1-13) so that all the $[CrCl^{2+}]$ terms appear on one side:

$$\frac{-d[CrCl^{2+}]/dt}{[CrCl^{2+}]} = k_1 + \frac{k_2}{[H^+]} \tag{1-14}$$

It turns out, as shown subsequently, that the quantity on the left-hand side of Eq. (1-14) is easily evaluated from the data, and we can equally well plot its logarithm (rather than log rate) against log $[H^+]$. That offers the advantage of "normalizing" every one of the rates to the same effective concentration of $CrCl^{2+}$, so that the variation of rate with $[H^+]$ reflects only its true effect. The indicated plot based on pub-

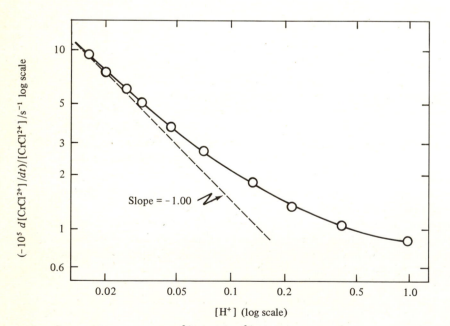

Figure 1-1 Plots of log $(-d[CrCl^{2+}]/dt/[CrCl^{2+}])$ versus log $[H^+]$ for the decomposition of $CrCl^{2+}$, Eq. (1-12).[2] The apparent reaction order with respect to $[H^+]$ is given by the slope of the curve, which approaches the limit -1 at low $[H^+]$ and 0 at high $[H^+]$.

lished kinetic data[2] for reaction (1-12) is shown in the form suggested by Eq. (1-14) in Fig. 1-1, and it has the expected form.

The above approach is used in practice as a means of formulating the correct algebraic relation between reaction rate and reagent concentrations. This issue will be considered in much more detail in Sec. 2-7, but it has been mentioned here because of its connection with the order of the reaction.

How is the rate of a reaction measured? Only in special circumstances is the reaction rate itself determined. One way to determine it is the so-called *method of initial rates*, which is based upon an accurate analysis of one product at a very early stage of reaction. During that interval, the reactant concentrations have changed but slightly, and that permits direct calculation of the rate as d[Product]$/dt$. Similar experiments with varying concentrations provide the data needed to formulate the rate law.

More often, one or more concentrations are determined throughout the timed course of a reaction. The data, in the form of concentrations as a function of time, are usually *not* processed in a way to yield a reaction rate. In the course of such a run the rate itself is constantly changing throughout; usually it declines steadily as the reactants are consumed. The rate might be evaluated from the tangent to a smooth curve drawn through the concentration-time data, although more often the kinetic data are evaluated after integration of the rate equation. These methods are the subject of Chaps. 2 to 4.

1-4 FACTORS INFLUENCING REACTION RATES

A number of variables are recognized to influence the rate of a reaction. The major factors are the following: (1) concentrations of reactants, products, and possibly catalysts, (2) physical conditions such as temperature and pressure, (3) the intensity of absorbed radiation, and, for reactions in solution, (4) properties of the solvent such as viscosity, dielectric constant, and ionic strength. Furthermore, the rate differences among a series of closely related compounds may provide a basis for making comparisons and mechanistic deductions beyond those possible from kinetic data for a single reaction.

Each of the variables will be considered in turn, and emphasis will be placed on the concentration variables especially. They are the variables which provide the most direct connection between chemical kinetics and reaction mechanism. The minimum goal of any kinetic study is to establish as well as possible the mechanism for the reaction in the sense in which the term was used earlier: to learn whether the reaction proceeds by a single pathway, whether a sequence of reactions occurs, whether equilibria precede the rate-limiting step, whether one or more reaction intermediates is involved. Included in this minimum set of questions is, of course, information about the rate-limiting step itself: Thus we seek to learn what reactant species are involved, to evaluate the rate constant(s) for the elementary reaction step(s), to establish the composition of the transition state(s) and of any intermediates. One hopes that the answers to these and related questions will prove unambiguous and complete, but that may or may not be the case.

If the determination of the reaction mechanism in the simplest instance is regarded as the resolution of the set of elementary reactions which constitute the pathway for the overall chemical process, the main variables are the set of rate-influencing concentrations. All of the systems we shall consider are *isothermal*; they are held at constant temperature by enclosing the reaction vessel in a thermostat.

Variation of temperature and pressure can provide additional information, particularly about the energetics and volume relations of reaction steps in the mechanism. The quantities deduced from the variations are collectively termed "activation parameters," and they include such values as the entropy of activation $\Delta S^{\ddagger}$ and the volume of activation $\Delta V^{\ddagger}$. If the reaction occurs simply as an elementary process, then such parameters are directly associated with the reaction itself. If the mechanism is more complicated, then the activation parameters will include contributions from steps preceding the rate-limiting step.

Some quantities associated with the rates and mechanisms of reactions are "determined"; they include the reaction rate under a given set of conditions, the rate constant, and the activation energy. Other quantities are "deduced" reasonably directly from the results of the experimental determinations: they include the elemental composition of the transition state and the nature of the rate-limiting reaction. Still other conclusions are "inferred" on grounds of varying soundness; examples are the polarity of the transition state, the degree of bond breaking and bond formation in the transition state, and the conformation of reaction intermediates.

1-5 PRACTICAL KINETICS

It is useful to consider briefly the experiments which must be conducted during the course of a kinetic study. Some preliminary work is usually in order, although much of the information one would thereby obtain may already be known. For example: What is the time scale over which the reaction occurs? The entire design of the experiments will differ if the reaction requires <1 s, 10 min, 10 h, or still longer. One might next consider what analytical technique(s) will be used to follow the reaction and whether the analysis will consist of a nondestructive monitoring of the mixture (perhaps a spectroscopic method such as ultraviolet-visible spectrophotometry or nuclear magnetic resonance) or will employ a method whereby individual samples (whether withdrawn aliquots or individual ampules) are analyzed. The choice of the method itself is governed by (1) the nature of the system and the reaction, (2) available instrumentation, (3) considerations of the concentration ranges which may need to be determined, (4) the time scale of the reaction of interest, and (5) the investigator's experience and familiarity with different techniques. Nearly every method of instrumental analysis has been applied to chemical kinetics at one time or another, and no single method can be said to dominate. The extent to which one can generalize is limited, but it is desirable to select methods which can provide analytical results to a precision no worse than 1 or 2 percent. When practicable the use of a second independent method for at least a few runs provides a desirable check; alternatively, one might employ the original method to determine a different reactant or product.

The choice of concentration ranges is still another parameter the investigator must select. Other considerations aside, it is useful to work with very dilute reagents to minimize effects of nonideality. Additional factors to be considered here, however, are the rates (time scale) of reaction at different concentrations, the other constituents present in the reaction mixture, and the sensitivity of the analytical method at different concentrations. Tied closely to this choice for reactions in solution is the question of solvent selection. The selection may be governed by the chemistry of the reaction. For example, water may be required in an enzymatic reaction having implications to living organisms, whereas strictly anhydrous conditions may be required by the instability of the reagents in a different system. Some reaction rates change markedly with solvent properties (polarity, dielectric constant, hydrogen bonding ability, viscosity), but others do not; indeed, the entire course of reaction may be altered, with a change in mechanism, possibly even a change in products, by a change in solvent. One cannot assume that the reaction is the same when such changes are made, and careful work requires a determination of product composition and stoichiometry under conditions as close as possible to those of the rate measurements.

Returning to the original point, we note that a brief survey in which some preliminary work is done is a great aid in the design of proper quantitative studies. When it comes time to launch the full-scale investigation, one must pay proper attention to reagent quality, solvent purification, and possible effects from such other factors as traces of moisture in organic solvents, the presence or exclusion of oxygen, catalytic impurities, and unanticipated salt effects. It is tempting then to proceed with the measurements themselves in a systematic manner, varying first one concentration regularly, then another, and so on. It is often preferable to randomize the variations, the better to detect any systematic change in rate over the course of the study arising from reagent decomposition or related sources. The set of experiments would usually be done at a constant temperature. The precision of the control needed to attain a given accuracy depends on the sensitivity of the reaction to temperature. The control required will be very different if a $10°$ increase in temperature causes the rate to increase by 20 percent, or by a factor of 5. Again, preliminary work is helpful in the design of experiments; one must consider not only the sensitivity of the thermostat controller but also the effect of a large heat of mixing. How accurately must one control temperature? The reader is aware from earlier studies (or see Sec. 6-1) that a rate constant depends upon temperature in an exponential fashion, as in the Arrhenius relation,

$$k = A \exp \frac{-E_a}{RT} \tag{1-15}$$

where the preexponential factor A and the activation energy E_a are adjustable parameters. From this we see that the error in k associated with a particular uncertainty in T is given by

$$\frac{\delta k}{k} = \frac{E_a}{RT} \frac{\delta T}{T} \tag{1-16}$$

For example, if one considers a reaction studied in the vicinity of room temperature with a so-called typical activation energy of $50,000$ J mol^{-1} (this corresponds to a reaction having a rate which doubles for a $10°$ temperature rise), the error in k associated with an uncertainty of $\pm 0.2°$ is

$$\frac{\delta k}{k} \times 100 = \frac{50{,}000 \text{ J mol}^{-1}}{8.31 \text{ J mol}^{-1} \text{ K}^{-1} \times 300 \text{ K}} \times \frac{0.2 \text{ K}}{300 \text{ K}} \times 100 = 1.4\% \qquad (1\text{-}17)$$

During the preliminary work it is also quite helpful to be alert for signs of unexpected complications. For example, do competing reactions occur or does the reaction appear to progress smoothly to completion? Are there indications that an intermediate may build up or that the product may not appear as rapidly as the reactant seems to disappear? Such complications are considered in Chaps. 3 and 4; an early recognition of them leads to more effective and efficient work.

REFERENCES

1. Lewis, E. S.: in A. Weissberger (ed.), "Investigations of Rates and Mechanisms of Reactions," 3d ed., vol. VI of the series "Techniques of Chemistry," Wiley-Interscience, 1974, p. 11.
2. Swaddle, T. W., and E. L. King: *Inorg. Chem.*, 4:532 (1965).

PROBLEMS

1-1 T. W. Newton, *J. Phys. Chem.*, 62:943 (1958), has shown that the major reaction between U(IV) and Pu(VI) is 2Pu(VI) + U(IV) = 2Pu(V) + U(VI) and that the rate of reaction can be expressed as

$$\frac{-d[\text{Pu(VI)}]}{dt} = k_a[\text{Pu(VI)}][\text{U(IV)}]$$

at constant $[\text{H}^+]$. Under certain conditions, $k_a = 2.2 \text{ dm}^3 \text{ mol}^{-1} \text{ s}^{-1}$.

(a) If one chose to reformulate the kinetic expression as

$$\frac{d[\text{U(VI)}]}{dt} = k_b[\text{Pu(VI)}][\text{U(IV)}]$$

what would be the value of k_b?

(b) If the rate-limiting step for the reaction is

$$\text{Pu(VI)} + \text{U(IV)} \xrightarrow{k_1} \text{Pu(V)} + \text{U(V)}$$

what is the second and final reaction? What is the numerical value of k_1?

(c) Devise a similar two-step mechanism different from the one in part b which would also be consistent with the kinetic data.

1-2 The reaction of vanadium (III) and chromium (II), $\text{V}^{3+} + \text{Cr}^{2+} = \text{V}^{2+} + \text{Cr}^{3+}$ occurs at a rate given by

$$\frac{-d[\text{V}^{3+}]}{dt} = k'[\text{V}^{3+}][\text{Cr}^{2+}]$$

with k' showing the following dependence upon $[\text{H}^+]$:

$$k' = \frac{q}{r + [\text{H}^+]}$$

State what the limiting order of the reaction is with respect to $[H^+]$ at low and high $[H^+]$. Depict a qualitative plot of log k' versus log $[H^+]$, and compare the result to Fig. 1-1.

BIBLIOGRAPHY

Bamford, C. H., and C. F. H. Tipper (eds.): "Comprehensive Chemical Kinetics," vols. 1–20, Elsevier Scientific Publishing Co., Inc., New York, 1969–1978.

Bell, R. P.: "The Proton in Chemistry," 2d ed., Cornell University Press, Ithaca, N.Y., 1973.

Benson, S. W.: "The Foundations of Chemical Kinetics," McGraw-Hill Book Company, New York, 1960.

Capellos, C., and B. H. J. Bielski: "Kinetic Systems: Mathematical Description of Chemical Kinetics in Solution," Wiley-Interscience, New York, 1972.

Frost, A. A., and R. G. Pearson: "Kinetics and Mechanism," 2d ed., John Wiley & Sons, Inc., New York, 1961.

Hammes, G. G.: "Investigations of Rates and Mechanisms of Reactions: Elementary Reaction Steps in Solution and Very Fast Reactions," vol. VI, part II of the series "Techniques of Chemistry," Wiley-Interscience, New York, 1974.

———: "Principles of Chemical Kinetics," Academic Press, New York, 1978.

Johnston, H. S.: "Gas Phase Reaction Rate Theory," Ronald Press, New York, 1966.

Jordan, P. C.: "Chemical Kinetics and Transport," Plenum Press, New York, 1979.

Laidler, K. J.: "Theories of Chemical Reaction Rates," McGraw-Hill Book Company, New York, 1969.

———: "Chemical Kinetics," 2d ed., McGraw-Hill Book Company, New York, 1965.

Lewis, E. S.: "Investigations of Rates and Mechanisms of Reactions: General Considerations and Reactions at Conventional Rates," vol. VI, part I of the series "Techniques of Chemistry," Wiley-Interscience, New York, 1974.

Skinner, G. B.: "Introduction to Chemical Kinetics," Academic Press, New York, 1974.

Wilkins, R. G.: "The Study of Kinetics and Mechanisms of Reactions of Transition Metal Complexes," Allyn & Bacon, Inc., Boston, 1974.

CHAPTER TWO

REACTIONS WITH A SIMPLE KINETIC FORM

Four subjects will be treated in this chapter: (1) kinetic equations for first- and second-order reactions and the analysis of kinetic data for those situations, (2) kinetic equations for other reaction orders, (3) methods of analysis of kinetic data obtained by instrumental methods, (4) determination or confirmation of the algebraic dependence of reaction rate upon concentrations.

Experiments in chemical kinetics usually do not consist of a direct evaluation of the reaction rate as such. Rather, most experimental techniques yield data giving concentration values as a function of time. The concentration of a reactant or product may be determined by direct chemical analysis or by measurement of a property of the system to which one or more of the substances in the reaction contributes. But kinetic data, to be readily interpreted in terms of reaction mechanism, generally require expression of the reaction rate as a function of concentration. Our first concern must, then, be for methods of transforming the concentration-time data into a rate-concentration form. This aspect of kinetic analysis is usually approached by the use of an integrated form of the rate equation.

A careful analysis at this stage is critical to a correct interpretation. In this chapter methods for conducting the analysis are applied to some situations in which the data follow a relatively straightforward kinetic equation.

2-1 FIRST-ORDER AND PSEUDO-FIRST-ORDER REACTIONS

One of the most widely encountered kinetic forms is that of the first-order rate equation. Consider a reaction $A \longrightarrow B$, proceeding irreversibly to completion, which follows the rate law

$$\frac{-d[A]}{dt} = k[A] \tag{2-1}$$

Equation (2-1) can be solved by integration between the time limits t_0, taken as zero, and time t and between concentration limits $[A]_0$ and $[A]_t$ (the latter often written simply as $[A]$). The integrated rate expression thus has the form

$$[A] = [A]_0 \exp(-kt) \quad \text{or} \quad \ln\frac{[A]}{[A]_0} = -kt \tag{2-2}$$

An alternative form of expression (2-2) results when the concentration of the product, B, is used as the variable instead of $[A]$. In the simplest case with stoichiometry $A \longrightarrow B$ the relations $[A] + [B] = [A]_0 = [B]_\infty$ and $[B]_0 = 0$, when substituted into Eq. (2-2), give the result

$$[B] = [A]_0 [1 - \exp(-kt)] \quad \text{or} \quad \ln\left(1 - \frac{[B]}{[A]_0}\right) = -kt \tag{2-3}$$

In the more general situation, a reaction with stoichiometry $aA = bB$ and a rate law written in the form $-(1/a)d[A]/dt = k[A]$, the equations are

$$\ln\frac{[A]}{[A]_0} = -akt \tag{2-2a}$$

$$\ln\left(1 - \frac{a[B]}{b[A]_0}\right) = -akt \tag{2-3a}$$

Before examining the analysis of experimental data, consider the phenomenon of *pseudo-first-order kinetics*, to which the same methods of data analysis will prove applicable. Imagine a reaction having a rate which depends on the concentrations of several substances. Also, imagine that conditions for the kinetic run are so selected that all but one of the concentrations are sufficiently high that, compared to the one reagent present at lower concentration, the others are effectively constant during the course of the run. This method of experimental design is referred to as the technique of *flooding*. Because only one concentration changes appreciably during the run, the effective kinetic order is reduced to the reaction order with respect to that one substance. If the order is unity, the reaction is said to follow pseudo-first-order kinetics in that particular run.

For example, consider again the reaction of BrO_3^- and Br^- in acidic solution, Eq. (1-9). Imagine that an experiment were carried out at relatively high concentrations of H^+ and Br^-, say, $[H^+]_0 = 1000 \times [BrO_3^-]_0$ and $[Br^-]_0 = 300 \times [BrO_3^-]_0$. The effective concentrations of H^+ and Br^- would be nearly constant, within the usual error limits of rate data. The rate equation would become:

$$\frac{-d[BrO_3^-]}{dt} = k[BrO_3^-][Br^-][H^+]^2 = k'[BrO_3^-] \tag{2-4}$$

The apparent rate constant k' is related to the "true" rate constant k (i.e., to one containing no further dependence on a concentration variable) and the particular $[Br^-]$ and $[H^+]$ in that experiment. The relation in this case is

$$k' = k[Br^-][H^+]^2 \tag{2-5}$$

The technique of flooding has served to reduce the fourth-order kinetic expression to a first-order one, with considerable mathematical simplification. Further examination of the method of flooding itself will be made in Sec. 2-7. Here our concern is with the kinetic analysis, and it should be noted that first- and pseudo-first-order kinetics are characterized by the same equations [Eqs. (2-2) and (2-3)].

The mathematical relations in Eq. (2-2) show that $[A]$ declines exponentially with time. According to the equation, a plot of ln $[A]$ against time will be linear with a slope of $-k$. The slope, and thus the rate constant, is independent of the initial concentration $[A]_0$. A further corollary for the first-order equation is that a particular fraction of the reaction occurs in a constant time that is characteristic of the given reaction and the conditions of the run in question but is independent of $[A]_0$. Consider the time required to complete exactly 50% reaction; substitution of $[A]_{1/2} = [A]_0/2$ into Eq. (2-2) gives the relation

$$t_{1/2} = \frac{\ln 2}{k} = \frac{0.693}{k} \tag{2-6}$$

We note that $t_{1/2}$, the *half-time* (or half-life) of the reaction, is independent of $[A]_0$ for a first-order rate equation. The quantity $1/k$ also has the dimensions of time; substitution into Eq. (2-2) reveals that it is the time required to reduce $[A]$ to $1/e$ of its initial value. That value of time, $1/k$, is known as the *mean reaction time*.

How does one know whether a reaction is properly described by a first-order kinetic equation? One proof comes from the linearity of the indicated plot of ln $[A]$ versus time, provided data are taken to at least 75 percent completion. Over a smaller interval, a plot that is really curved may appear to be straight within experimental error. A second and better substantiation of first-order kinetics is the constancy of the rate constant k derived from experiments having different values of $[A]_0$. If possible, a 10-fold variation of $[A]_0$, at least, is desirable.

The first-order rate constant can be evaluated in several ways from the kinetic data:

1. Every point is referred to the initial point, and an average of the individual values is taken. Each value is given by

$$k = \frac{1}{t} \ln \frac{[A]_0}{[A]_t} \tag{2-7}$$

2. Adjacent pairs of points are used to compute k, and the values are then averaged. The expression for k is

$$k = \frac{1}{t' - t''} \ln \frac{[A]''}{[A]'} \tag{2-8}$$

3. The data are fit by graphical or numerical techniques to the integrated rate law; that is, k is given by the slope of the plot of ln [A] versus time:

$$\ln [A] = \ln [A]_0 - kt \qquad (2\text{-}9)$$

Method 1 weights the starting value heavily, and method 2 involves small differences. Indeed, if the time interval between successive points is identical, method 2 is equivalent to computing k from the first point and the last, ignoring all the rest (see Prob. 2-1). Method 3 is preferable in most instances. The concentrations can be multiplied by a weighting factor[1] to allow for data of unusually high precision (note that the lack of weighting does not necessarily amount to giving equal weight to each point and may not be appropriate) or for points having very different degrees of precision.

Since the rate constant does not depend upon $[A]_0$, neither the initial concentration nor the exact instant at which the reaction began need be known accurately. That need not be taken as rationalization for the adoption of poor experimental techniques, but it is an important aspect of first-order kinetics.

An illustrative reaction is the hydrolysis of triphenylmethyl chloride,

$$Ph_3CCl + H_2O = Ph_3COH + H^+ + Cl^- \qquad (2\text{-}10)$$

The reaction can be followed by titration of the acid liberated during the course of the experiment. Sample kinetic data are given in Table 2-1. Since this particular set of data[2] gives the concentration of a reaction product, that of the reactant is obtained by difference. The plot of the data is conveniently made by plotting either ln $[Ph_3CCl]$ versus time or simply $[Ph_3CCl]$ versus time by using semilogarithmic graph paper; both ordinates are shown in Fig. 2-1. From the slope of the straight line, the rate constant is calculated to be $4.3_4 \times 10^{-3}$ s^{-1}. This corresponds to $t_{1/2} = 160$ s.

Table 2-1 Kinetic data[a,b] for the hydrolysis of triphenylmethyl chloride, Eq. (2-10)

Time/s	Concentrations/M	
	$10^3 [H^+]_t$	$10^3 [Ph_3CCl]_t^c$
0	—	1.44
18	0.104	1.34
57	0.312	1.13
93	0.484	0.96
171	0.757	0.68
298	1.04	0.40
448	1.23	0.21
508	1.28	0.16
1800("∞")	1.44	—

[a]Data from Ref. 2.
[b]Conditions: $[Ph_3CCl]_0 = 1.44 \times 10^{-3}$ M, 15% water–85% acetone at $-34.2°$C.
[c]Calculated as $[Ph_3CCl]_0 - [H^+]_t$.

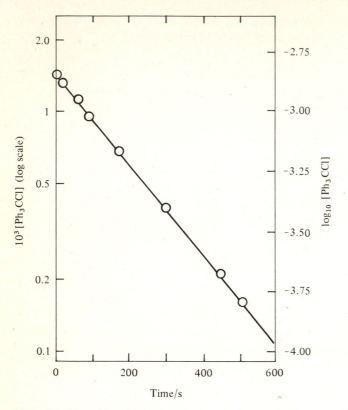

Figure 2-1 Plot of the first-order kinetic data for hydrolysis to triphenylmethyl chloride given in Table 2-1. The slope of this line gives the value of $-k$; the value in this run is $k = 4.3_4 \times 10^{-3}$ s^{-1}. Data from Ref. 2. The right ordinate shows the logarithms of the concentration, whereas the left ordinate uses the logarithmic scale on which the concentrations are plotted directly.

2-2 SECOND-ORDER KINETICS

Second-order kinetics arise when the rate is proportional to the square of the concentration of a single reagent and also when it is proportional to the product of the concentrations of two reagents. In this author's experience, the former is relatively rare in practice and the latter is exceptionally common.

The first case is described by the rate law

$$\frac{-d[A]}{dt} = k[A]^2 \qquad (2\text{-}11)$$

After integration between the usual limits, the resulting expression is

$$\frac{1}{[A]} = \frac{1}{[A]_0} + kt \qquad (2\text{-}12)$$

Thus a plot of $1/[A]$ versus time will be linear with slope k. The graph, by the plotting of reciprocal concentrations, tends to emphasize the least accurate points near the end of the run. An alternative procedure is based upon the rearranged equation:

$$[A] = [A]_0 - k[A]_0 \, t[A] \tag{2-13}$$

A plot of $[A]$ against the product $t[A]$ will be linear with a slope of $-k[A]_0$ if the second-order rate law of Eq. (2-11) is correct. As an illustration of the plots, consider kinetic data[3] on the disproportionation of p-toluenesulfinic acid in aqueous, acidic acetic acid.

$$3\,ArSO_2H = ArSO_2SAr + ArSO_3H + H_2O$$

$$(Ar = p\text{-}CH_3C_6H_4\text{—})$$

The reaction follows a second-order rate law. Data from one experiment are shown in Table 2-2, and the plots according to Eqs. (2-12) and (2-13) are shown in Fig. 2-2. Based on these calculations, the rate constant is 2.2×10^{-3} dm^3 mol^{-1} s^{-1}.

The rate law for "mixed-second-order" kinetics,

$$\frac{-d[A]}{dt} = k[A][B] \tag{2-14}$$

is one of the most widely encountered forms. Consider first the situation in which A and B react with $1:1$ stoichiometry, using the notation that A is the limiting reagent. From the stoichiometry, the concentration of B, the excess reagent, is given by

$$[B] = [B]_0 - [A]_0 + [A] = \Delta_0 + [A] = [B]_\infty + [A]$$

The differential equation to be solved is thus

$$\frac{-d[A]}{[A](\Delta_0 + [A])} = k \, dt$$

Table 2-2 Kinetic data[a,b] for the disproportionation of
p-toluenesulfinic acid

Time/s	$[ArSO_2H]/M$	$[ArSO_2H]^{-1}/M^{-1}$	$t[ArSO_2H]/Ms$
0	0.100	10.00	0
900	0.0863	11.59	77.7
1,800	0.0752	13.30	135.4
2,700	0.0640	15.63	172.8
3,600	0.0568	17.61	204.5
7,200	0.0387	25.84	278.6
10,800	0.0297	33.67	320.8
18,000	0.0196	51.02	352.8

[a]At 70°C in acetic acid-0.56 M H_2O; 1 M H_2SO_4.
[b]Data from Ref. 3.

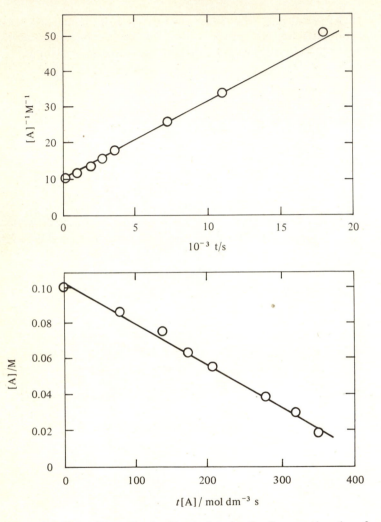

Figure 2-2 Plots of second-order kinetic data for the disproportionation of p-toluenesulfinic acid. The upper plot shows $[A]^{-1}$ versus time, according to Eq. (2-12), and the lower plot is that of $[A]$ versus $t[A]$, according to Eq. (2-13). $A = p\text{-CH}_3\text{C}_6\text{H}_4\text{SO}_2\text{H}$. Data from Table 2-2 and Ref. 3.

This equation can be integrated by partial fractions or more simply from a table of integrals, which gives

$$\int \frac{dX}{X(C+X)} = -\frac{1}{C} \ln \frac{C+X}{X}$$

With the usual limits of integration, the solution is

$$\ln \frac{[B]}{[A]} = \ln \frac{[B]_0}{[A]_0} + ([B]_0 - [A]_0)kt \qquad (2\text{-}15)$$

Table 2-3 Kinetic data[a,b] for a reaction following mixed second-order kinetics [reaction (2-18)]

Time/s	$[RN(CH_3)_2]$/M	$[OH^-]$/M	$\dfrac{[OH^-]}{[RN(CH_3)_2]}$
0	0.0500	0.199	3.98
135	0.0413	0.190	4.60
380	0.0365	0.186	5.10
610	0.0325	0.182	5.60
945	0.0282	0.177	6.28
1880	0.0187	0.168	8.98

[a]From Ref. 4.
[b]Conditions: 50% aqueous ethanol at 50°C.

In the general case of a reaction occurring with the stoichiometry $a\mathrm{A} + b\mathrm{B} = \text{prod-}$ ucts, at a rate given by

$$\frac{-d[A]}{dt} = ak[A][B] \tag{2-16}$$

the integrated equation is

$$\ln\frac{[B]}{[A]} = \ln\frac{[B]_0}{[A]_0} + (a[B]_0 - b[A]_0)kt \tag{2-17}$$

By way of illustration, consider the kinetics of the alkaline hydrolysis of p-nitrosodimethylaniline:

$$p\text{-ON-}C_6H_4N(CH_3)_2 + OH^- \longrightarrow p\text{-ON-}C_6H_4O^- + (CH_3)_2NH \tag{2-18}$$

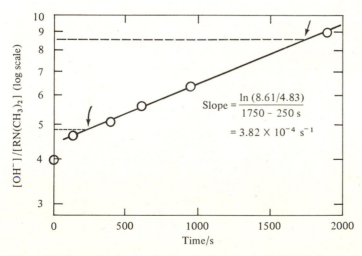

Figure 2-3 Second-order kinetic plot for the data in Table 2-3 for reaction (2-18). The plot is of log ([B]/[A])—actually [B]/[A] on a logarithmic scale—against time, according to Eq. (2-15). The value of the rate constant is slope/([B]$_0$ − [A]$_0$).

Kinetic data[4] for a sample run are summarized in Table 2-3. The plot of log ([OH⁻]/[aniline]) can be facilitated by use of semilogarithmic graph paper. Figure 2-3 shows the plot. The slope of the straight line given in terms of natural logarithms is 3.85×10^{-4} s⁻¹. The value of k according to Eq. (2-15) is the quotient of the slope and the difference in reactant concentrations:

$$k = \frac{3.85 \times 10^{-4} \text{ s}^{-1}}{(0.199 - 0.050) \text{ M}} = 2.58 \times 10^{-3} \text{ M}^{-1} \text{ s}^{-1}$$

The calculation of the rate constant depends upon the accuracy with which $[B]_0 - [A]_0$ is known. The closer the concentrations, the less accuracy remains in their difference. When $[B]_0$ becomes quite close to $[A]_0$, further complications arise, as discussed in the next section.

2-3 EQUIVALENT CONCENTRATIONS IN SECOND-ORDER KINETICS

Consider a reaction $aA + bB =$ products that might follow any of these three rate expressions for $-d[A]/dt$:

$$(1) \ k_1[A]^2 \qquad (2) \ k_2[A][B] \qquad (3) \ k_3[B]^2 \qquad\qquad (2\text{-}19)$$

What is learned in each case in an experiment with stoichiometric (equivalent) concentrations $(1/a)[A]_0 = (1/b)[B]_0$? Under those conditions it is easy to see that the three expressions are equivalent. In all three cases, therefore, a plot of $1/[A]$ or $1/[B]$ against time would be linear, because the kinetics would follow Eq. (2-12). It is easy to conclude that this experiment is not a very informative one, because it does not differentiate among the three forms shown. That is to say, the seeming mathematical simplification comes at the expense of the information obtained from the single kinetic experiment having nonequivalent concentrations.

Since the three differential equations in (2-19) reduce to the same form in an experiment having stoichiometric concentrations, it follows that the integrated rate laws for the two cases should become equivalent; i.e., Eq. (2-15) should reduce to Eq. (2-12) in the limit $[B]_0 = [A]_0$. The expectation is not immediately realized, however, since Eq. (2-15) becomes indeterminate, 0/0, with the substitution $[B]_0 = [A]_0$. Some treatments stop at this point, instructing the reader to make the substitution before integration and thereby reducing case 2 to case 1.

Alternative approaches are to use L'Hospital's rule[†] or to use a series expansion[5a] to surmount the difficulty. Based upon the first method[5b], the rate constant is given by the following limit:

[†]L'Hospital's rule is $\lim\limits_{x \to a} \dfrac{f(x)}{g(x)} = \lim\limits_{x \to a} \dfrac{f'(x)}{g'(x)}$

$$k = \lim_{[B]_0 \to [A]_0} \left\{ \frac{(d/d[B]_0)\{\ln [A]_0 + \ln ([B]_0 - [A]_0 + [A]) - \ln [A] - \ln [B]_0\}}{(d/d[B]_0)([B]_0 t - [A]_0 t)} \right\}$$

$$= \lim_{[B]_0 \to [A]_0} \left\{ \frac{([B]_0 - [A]_0 + [A])^{-1} - [B]_0^{-1}}{t} \right\} = \frac{1}{t}\left(\frac{1}{[A]} - \frac{1}{[A]_0} \right)$$

which correctly transforms to Eq. (2-12) and constitutes a proof of the validity of this point. The second proof[5a] is based upon the expansion of $\ln (1 + y)$ for $-1 < y < 1$ in a convergent series:

$$\ln (1 + y) = y - \frac{y^2}{2} + \frac{y^3}{3} - \cdots \tag{2-20}$$

The integrated rate law of Eq. (2-15) can be rearranged to the form

$$kt = \frac{1}{\Delta_0}\left(\ln \frac{[A]_0}{[B]_0} + \ln \frac{[B]}{[A]} \right)$$

$$= \frac{1}{\Delta_0}\left[-\ln \left(1 + \frac{\Delta_0}{[A]_0} \right) + \ln \left(1 + \frac{\Delta_0}{[A]} \right) \right]$$

Each logarithmic term is then expanded as indicated in Eq. (2-20). Powers beyond the first for $y \ll 1$ are dropped, giving

$$kt = \frac{1}{\Delta_0}\left(-\frac{\Delta_0}{[A]_0} + \frac{\Delta_0}{[A]} \right)$$

which is again Eq. (2-12).

Kinetic runs with equivalent concentrations are less desirable experiments in any case, and they are probably best avoided unless dictated by special circumstances. If there is cause to consider such experiments, then there arises the practical problem of attaining *exactly* identical concentrations. If $[A]_0$ and $[B]_0$ are nearly but not exactly equal, Eq. (2-15) is of little use, since the difference Δ_0 is highly inaccurate with $[B]_0$ only slightly different from $[A]_0$. The difficulty can be resolved by using average concentrations.

$$\overline{[A]} = \overline{[B]} = [A] + \frac{\Delta_0}{2} = [B] - \frac{\Delta_0}{2} \tag{2-21}$$

Then the rate law becomes

$$\frac{-d[A]}{dt} = \frac{-d\overline{[A]}}{dt} = k\left(\overline{[A]} - \frac{\Delta_0}{2} \right)\left(\overline{[A]} + \frac{\Delta_0}{2} \right) \tag{2-22}$$

$$\frac{-d\overline{[A]}}{dt} = k\left(\overline{[A]}^2 - \frac{\Delta_0^2}{4} \right) \tag{2-23}$$

For small values of Δ_0, $-d\overline{[A]}/dt \simeq k\overline{[A]}^2$ and Eq. (2-12) or (2-13) with $\overline{[A]}$ replacing $[A]$ may be used.

2-4 USE OF PHYSICAL PROPERTIES IN FIRST- AND SECOND-ORDER KINETICS

Instrumental determinations have largely supplanted direct analyses for the purpose of following the progress of a reaction. Thus a property such as the volume of the liquid phase (dilatometry), the angle of optical rotation, the absorbance at a fixed wavelength (spectrophotometry), the area of a nmr peak, the conductance of a solution, or the pressure of a gas mixture is followed with time. In the simplest situation, say a first-order reaction in which only the reactant A makes a contribution to this property P, it is easy to see that the fractional change in P is proportional to [A] and may be substituted for it in Eq. (2-2).

It is also true, although not so easily seen, that the same result also holds when the product contributes to the value of P. The development of this case for the irreversible first-order reaction $A \longrightarrow B$ is as follows. If p_A and p_B represent the proportionality constants between concentration and property (i.e., the partial molar volumes in dilatometry, molar absorptivities, or extinction coefficients in spectrophotometry, etc.) and C is a constant contribution from the solvent or other inert components, then the value of P at any time t is given by

$$P_t = p_A[A] + p_B[B] + C \tag{2-24}$$

Substitution of the first-order equations gives

$$P_t = p_A[A]_0 \exp(-kt) + p_B[A]_0[1 - \exp(-kt)] + C \tag{2-25}$$

The initial and final instrument readings are

$$P_0 = p_A[A]_0 + C$$

$$P_\infty = p_B[A]_0 + C$$

Substitution of those relations into Eq. (2-25) yields, upon rearrangement,

$$\frac{P_t - P_\infty}{P_0 - P_\infty} = \exp(-kt) \tag{2-26}$$

Thus a plot of $\ln(P_t - P_\infty)$, or $\ln(P_\infty - P_t)$ if the values of P increase with time, against time is a straight line of slope $-k$. No assumption concerning the relative values of p_A and p_B are made, and Eq. (2-26) is generally valid for irreversible reactions following a first-order rate law.

The property P has been assumed to be linearly related to concentration; in spectrophotometry, for example, each reactant and product obeys Beer's law over the concentration range of interest. That should be checked; if it is not validated, the method may still be reliable with proper calibration. In such a case, however, the nonlinearity may signal important interactions among species which must be resolved. We also recognized that a property such as conductance gives resistivity values which bear a reciprocal rather than direct proportionality to concentration, but that is easily allowed for in the context of the present discussion.

Consider now the general case, that of reactions other than first-order. The relation between the observed property and the concentrations in the reaction is independent of the form of the kinetic equations. For any general reaction,

$$a\text{A} + b\text{B} + \cdots = v\text{V} + w\text{W} + \cdots \tag{2-27}$$

which proceeds to completion, the concentration of A, the limiting reagent, at any time t is given by

$$[\text{A}]_t = [\text{A}]_0 \frac{P_t - P_\infty}{P_0 - P_\infty} \tag{2-28}$$

where the values of property P are evaluated at the start and end of the reaction, P_0 and P_∞, and at time t. Perhaps Eq. (2-28) can be seen to be valid on an intuitive basis, but it follows equally well from the following derivation. At time t the value of the property P is

$$P_t = p_\text{A}[\text{A}] + p_\text{B}[\text{B}] + p_\text{V}[\text{V}] + p_\text{W}[\text{W}] + \cdots + C \tag{2-29}$$

Values of initial and final readings are

$$P_0 = p_\text{A}[\text{A}]_0 + p_\text{B}[\text{B}]_0 + C \tag{2-30}$$

$$P_\infty = p_\text{B}\left([\text{B}]_0 - \frac{b}{a}[\text{A}]_0\right) + p_\text{V}\frac{v}{a}[\text{A}]_0 + p_\text{W}\frac{w}{a}[\text{A}]_0 + \cdots + C \tag{2-31}$$

The value of P_t, after substitution of the stoichiometric relations, becomes

$$P_t = p_\text{A}[\text{A}] + p_\text{B}\left([\text{B}]_0 - \frac{b}{a}[\text{A}]_0 + \frac{b}{a}[\text{A}]\right) + p_\text{V}\frac{v}{a}([\text{A}]_0 - [\text{A}]) + \cdots + C \tag{2-32}$$

Combination of Eqs. (2-30) to (2-32) affords Eq. (2-28), which has thus been shown to be valid in general for the case of a single stoichiometric net reaction having constituents which contribute linearly to a property P, *provided the reaction proceeds to completion.*[†]

For the three rate laws thus far considered the expressions in terms of property P are:

First order, $-d[\text{A}]/dt = k[\text{A}]$

$$\ln \frac{P_t - P_\infty}{P_0 - P_\infty} = -kt \tag{2-33}$$

Second order, $-d[\text{A}]/dt = k[\text{A}]^2$

$$P_t = P_0 - k[\text{A}]_0 t(P_t - P_\infty) \tag{2-34}$$

[†]If the reaction is not complete and an equilibrium is attained, the general expression for the concentration of the limiting reagent is

$$[\text{A}]_t = [\text{A}]_\infty + ([\text{A}]_0 - [\text{A}]_\infty)\frac{P_t - P_\infty}{P_0 - P_\infty} \tag{2-28a}$$

Second order, $-d[A]/dt = k[A][B]$

$$\ln\left(1 + \frac{\Delta_0}{[A]_0}\frac{P_0 - P_\infty}{P_t - P_\infty}\right) = \ln\frac{[B]_0}{[A]_0} + \Delta_0 kt \tag{2-35}$$

Proper use of these relations requires an accurate value of P_∞. Any small error in P_∞ will result in kinetic plots which show systematic upward or downward curvature near the end of the run. The deviations are systematic rather than random because they arise from repeated use of the same incorrect value. As the end of the run nears, the value of P_t approaches P_∞, and the small difference is subject not only to random errors in P_t but also to any systematic error in P_∞. A small error in P_∞ is one of the most frequent sources of curvature in plots of kinetic data.

How far must a reaction be followed to obtain a reliable value of P_∞? The answer can be visualized when expressed in terms of the number of half-time periods $t'/t_{1/2}$, where t' is the time one elects to call adequately close to the final time. For a reaction second-order in $[A]$, $t_{1/2} = 1/k[A]_0$; each half-time period is double the preceding one. We use the first $t_{1/2}$ for a second-order reaction to express the results, which are summarized as follows:

	Percent completed		
	90%	99%	99.9%
$t'/t_{1/2}$ for first-order	3.32	6.64	9.97
$t'/t_{1/2}$ for second-order	9.0	99	999

It is clear from this comparison that the rate of change in a second-order reaction becomes so slow near completion that waiting for even 99 percent completion may be quite tedious. Hence a procedure is often adopted whereby a value of P_∞ calculated from the known properties of the products is used after a lapse of the requisite length of time in only a few runs to check the reliability of the calculated P_∞. Also, it is quite common now to use a digital computer to fit experimental data to equations such as these, with the provision that both P_∞ and k are to be treated as adjustable parameters. This will provide the least-squares values and standard deviations of P_∞ and k and as many statistical measures of the goodness of the fit as one desires.[6] The programs are rapid and inexpensive. One must, of course, check that the computer-fit value of P_∞ is within experimental error of the correct or expected value.

2-5 METHODS WHEN THE INFINITY READING OR "END POINT" IS UNKNOWN

Kinetic data in which the final instrument reading P_∞ is unreliable or unavailable are sometimes taken. That might happen because an excessive length of time would be required, because a slow secondary reaction sets in, or because of slow baseline drift. In certain circumstances, especially when the reaction follows first-order kinetics, the data can be treated quite satisfactorily without introduction of the value of P_∞.

Two methods for first-order-kinetic data, attributed to Guggenheim[7] and to Kezdy[8] and Swinbourne[9], are available. Consider first-order kinetic data at time t and at a time τ later; the value of P at each time is

$$P_t = (P_0 - P_\infty) \exp(-kt) + P_\infty$$

$$P_{t+\tau} = (P_0 - P_\infty) \exp[-k(t + \tau)] + P_\infty$$

In the Guggenheim treatment, the two equations are subtracted:

$$P_t - P_{t+\tau} = (P_0 - P_\infty)[1 - \exp(-k\tau)] \exp(-kt) \qquad (2\text{-}36)$$

All quantities on the right of Eq. (2-36) except the last are time-independent. A plot of $\ln(P_t - P_{t+\tau})$ versus time is linear with slope $-k$. Each point appearing in the plot depends upon two readings, as before, but now upon two different readings. For best accuracy τ is chosen as two to three half-times.

In the Kezdy-Swinbourne treatment two spaced values are divided:

$$\frac{P_t - P_\infty}{P_{t+\tau} - P_\infty} = \exp(k\tau)$$

Solving for P_t,

$$P_t = P_\infty[\exp(k\tau) - 1] + P_{t+\tau} \exp(k\tau) \qquad (2\text{-}37)$$

Note that the first term on the right is time-independent. Hence a plot of P_t versus $P_{t+\tau}$ will be linear with a slope of $\exp(k\tau)$. The rate constant is then calculated from the slope by using the value of τ chosen in originally dividing the data set into two parts. A value of τ between $t_{1/2}$ and $1.5t_{1/2}$ provides greatest accuracy. Since at the end point $P_t = P_{t+\tau} = P_\infty$, the intersection of the line through the data points with the $45°$ line gives the value of P_∞. One can thus extrapolate to the end-point reading quite readily. In fact, some workers will use the Kezdy-Swinbourne method only to estimate P_∞, and then use the value in the complete expression of Eq. (2-33) to evaluate the rate constant.

Both methods in which P_∞ is not explicitly used are of comparable accuracy, although neither provides as good a treatment as the direct use of P_∞ in Eq. (2-33) when a reliable value is known. The Kezdy-Swinbourne method is perhaps the more convenient, since the original data are plotted directly without first computing differences and logarithms. Readings toward the end of the reaction are "telescoped" on the Kezdy-Swinbourne graph and thus tend to contribute less than the points at the beginning of the run. This is a desirable feature; the value of P is changing more rapidly at the start of the run, which gives a more accurate measure of the rate.

To illustrate the two methods we shall use published data[10] for the first-order decomposition of "diacetone alcohol" (2-methyl-2-pentanol-4-one) in the presence of amines:

$$(CH_3)_2C(OH)CH_2C(O)CH_3 \longrightarrow 2(CH_3)_2CO \qquad (2\text{-}38)$$

Data for a typical run are given in Table 2-4; they are based on dilatometer readings. Figure 2-4 shows the Kezdy-Swinbourne method applied to the data. The slope

Table 2-4 Kinetic data[a,b] for reaction (2-38)

Time/s	Dilatometer readings		Guggenheim method	P_∞ method[c]	
	v_t	$v_{t+7\,h}$	$v_t - v_{t+7\,h}$	$v - v_\infty$	$v_{t+7\,h} - v_\infty$
0	83.46	46.94	36.52	66.66	30.14
900	81.52	46.08	35.44	64.72	29.28
1,800	79.66	45.26	34.40	62.86	28.46
2,700	77.89	44.48	33.41	61.09	27.68
3,900	75.60	43.42	32.18	58.80	26.62
5,100	73.40	42.42	30.98	56.60	25.62
8,100	68.32	40.05	28.27	51.52	23.25
9,300	66.42	39.25	27.17	49.62	22.45
10,500	64.64	38.42	26.22	47.84	21.62
11,700	62.88	37.62	25.26	46.08	20.82
13,500	60.38	36.50	23.88	43.58	19.70
14,700	58.80	35.80	23.00	42.00	19.00
15,900	57.25	35.05	22.20	40.45	18.25
17,100	55.74	34.39	21.35	38.94	17.59
18,900	53.58	33.42	20.16	36.78	16.62

[a]Data from Ref. 10.
[b]Conditions: 0.080 M diacetone alcohol at $T = 18.05°C$.
[c]Using $v_\infty = 16.80$ as estimated from the Kezdy-Swinbourne plot shown in Fig. 2-4.

of the line is 2.21, and based on the value of $\tau = 7$ h, or 25,200 s, this yields the first-order rate constant

$$k = \frac{\ln \text{slope}}{\tau} = \frac{\ln 2.21}{25,200 \text{ s}} = 3.15 \times 10^{-5} \text{ s}^{-1}$$

The Guggenheim plot is shown in Fig. 2-5, and it gives $k = 3.17 \times 10^{-5}$ s^{-1} from this graphical treatment. Also shown in Fig. 2-5 is a plot of the same data using the extrapolated end-point reading from the Kezdy-Swinbourne plot; the method gives $k = 3.15 \times 10^{-5}$ s^{-1} based on the graph shown.

Both the Guggenheim and the Kezdy-Swinbourne plots must be linear, within experimental error, if the reaction follows first-order kinetics. Yet the linearity of these plots, unlike that of the plot of log $(P_t - P_\infty)$ against time, does not constitute proof of first-order kinetics. Other kinetic equations also may lead to linear plots of either function. The assertion is illustrated by the example given in Prob. 2-7.

Hartley[11] has devised a procedure for unequal time intervals, although its application appears tedious. Moore[12] has also produced a computational method which is essentially a routine for determining both k and P_∞, and other such computer methods[6] were mentioned earlier in this chapter. A method[13] for second-order reactions analogous to the Guggenheim and Kezdy-Swinbourne procedures will now be considered. The method is applicable to a rate law which is second-order in a single component, $-d[A]/dt = k[A]^2$. It also applies to the mixed-second-order expression $-d[A]/dt = k[A][B]$, but only when the reactants and products are present at equivalent concen-

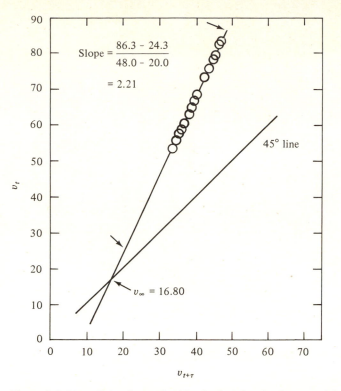

Figure 2-4 Plot of the first-order kinetic data for the decomposition of diacetone alcohol using the Kezdy-Swinbourne method. Data from Table 2-4 and Ref. 10. The value of the time increment τ is 7 hr or 25,200 s. The intersection of the line through the data points with the 45° line gives an extrapolated value of the end-point reading.

trations. The basic equation was given earlier:

$$P_t = P_0 - k[A]_0 t (P_t - P_\infty) \qquad (2\text{-}34)$$

If a similar expression is written for the value of P at a time τ later and the two equations are subtracted, the following expression is obtained after rearrangement:

$$P_t - P_{t+\tau} = k[A]_0 [\tau P_{t+\tau} - t(P_t - P_{t+\tau})] - k[A]_0 \tau P_\infty \qquad (2\text{-}39)$$

Since the last term of Eq. (2-39) is constant (time-independent), the rate constant can be evaluated by a plot of $P_t - P_{t+\tau}$ versus the quantity $\tau P_{t+\tau} - t(P_t - P_{t+\tau})$. This plot should be linear with a slope equal to $k[A]_0$. An illustration of the method uses published data[14] for the reaction of plutonium(VI) and uranium(IV):

$$2Pu(VI) + U(IV) = 2Pu(V) + U(VI)$$

which follows the rate law

$$\frac{-d[Pu(VI)]}{dt} = k[Pu(VI)][U(IV)]$$

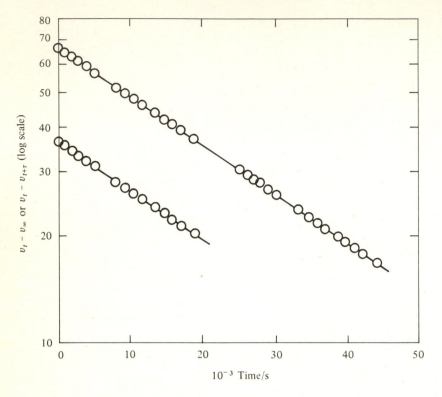

Figure 2-5 Semi-logarithmic plots of the kinetic data for the decomposition of diacetone alcohol using data from Table 2-4 and Ref. 10.

Upper line: A plot of $v_t - v_\infty$ on a logarithmic scale using $v_\infty = 16.80$ as derived from the Kezdy-Swinbourne extrapolation, Fig. 2-4.

Lower line: Kinetic plot of the same data using the Guggenheim method with $\tau = 7$ hr or 25,200 s.

In an experiment with equivalent concentrations (owing to the $2:1$ stoichiometry of the reaction, this refers to a run having $[Pu(VI)]_0 = 2[U(IV)]_0$), the rate equations can be transformed for this run only to a mathematically equivalent form,

$$\frac{-d[Pu(VI)]}{dt} = \frac{k}{2}[Pu(VI)]^2$$

since $[U(IV)]_t = [Pu(VI)]_t/2$.

The experimental data were obtained spectrophotometrically, and the data for one run conforming to the conditions are shown in Table 2-5. The graph is given in Fig. 2-6. The unweighted least-squares line through the points has an intercept of -0.0160 ± 0.0066 and a slope of $(4.56 \pm 0.11) \times 10^{-3}$ s^{-1}, from which the rate constant is given by

$$k = \frac{2 \times \text{slope}}{[A]_0} = \frac{2 \times 4.56 \times 10^{-3} \text{ s}^{-1}}{1.802 \times 10^{-4} \text{ mol/dm}^3}$$

Table 2-5 Kinetic data^a for the second-order reaction of Pu(VI) and U(IV) analyzed by a time-lag method ($\tau = 300$ s)

Time/s	D_t	D_{t+300}	$D_t - D_{t+300}$	$[\tau D_{t+300} - t(D_t - D_{t+300})]/s^{-1}$
60	0.757	0.370	0.387	87.8
120	0.625	0.339	0.286	67.4
180	0.533	0.308	0.225	51.9
240	0.464	0.286	0.178	43.1
300	0.412	0.265	0.147	35.4

^{a}Data from Ref. 14, referring to values of absorbance D from a run at 25.0°C, 2.0 mol dm^{-3} ionic strength, 0.150 mol dm^{-3} H$^+$, with [Pu(VI)]$_0$ = 1.803 × 10^{-4} mol dm^{-3} and [U(IV)]$_0$ = 0.902 × 10^{-4} mol dm^{-3}.

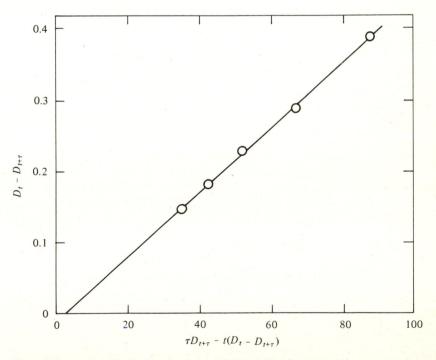

Figure 2-6 A plot of the kinetic data of Table 2-5, referring to the second-order reaction of plutonium(VI) and uranium(IV). The plot given is that suggested by Eq. (2-39), showing $D_t - D_{t+300}$ versus $300 D_{t+300} - t(D_t - D_{t+300})$, where D is the absorbance reading at the time indicated by the subscript.

or $k = 50.6 \pm 1.2$ dm^3 mol^{-1} s^{-1}. The intercept gives the end-point reading as $D_\infty = 0.012 \pm 0.005$.

2-6 OTHER SIMPLE KINETIC FORMS

The reader will experience no difficulty in deriving integrated rate laws for other single-term rate expressions, a number of which are summarized in Table 2-6 for sake of convenient reference.

2-7 RATE EXPRESSIONS AND THE METHOD OF FLOODING

Much time could be spent integrating all conceivable rate equations of the form

$$\frac{-d[A]}{dt} = k[A]^x [B]^y [C]^z$$

The time does not appear to be well spent, because experimental conditions are generally selected to simplify the form of the rate law. In practice, the technique of flooding would most likely be used to reduce the experimental reaction order to a value lower than $x + y + z$. That is to say, many of the substances which influence the reaction rate would be added independently just for the purpose of simplifying the kinetic form. As such, these reagents would remain at a concentration which is essentially constant during the experiment.

How high must a reagent concentration be, compared to that of the limiting reagent, to be considered to remain "essentially constant"? How much error is intro-

Table 2-6 Summary of integrated rate expressions for selected single-term rate equations[a]

$-d[A]/dt =$	Integrated rate law
1. k	$[A] = [A]_0 - kt$
2. $k[A]^{1/2}$	$[A]^{1/2} = [A]_0^{1/2} - (k/2)t$
3. $k[A]$	$\ln([A]/[A]_0) = -kt$
4. $k[A]^{3/2}$	$[A]^{1/2} = [A]_0^{1/2} - (k[A]_0^{1/2}/2)([A]^{1/2}t)$
5. $k[A]^2$	$[A] = [A]_0 - (k[A]_0)([A]t)$
6. $k[A]^n$ $(n \neq 1)$	$[A]^{n-1} = [A]_0^{n-1} - (k[A]_0^{n-1})(n-1)([A]^{n-1}t)$
7. $k[B]$	$\ln(\Delta_0 + [A]) = \ln[B] = \ln[B]_0 - kt$
8. $k[A][B]$	$\ln([B]/[A]) = \ln([B]_0/[A]_0) + k\Delta_0 t$
9. $k[A][C]$	$\ln\left(\dfrac{[A]}{[A]_0 + [C]_0 - [A]} \dfrac{[C]_0}{[A]_0}\right) = -k([A]_0 + [C]_0)t$

[a]A and B react with $1:1$ stoichiometry; $[B]_0 > [A]_0$. C is used to represent a product of the reaction; $[C] = [A]_0 - [A] + [C]_0$. Reaction 9 is autocatalytic. $\Delta_0 = [B]_0 - [A]_0$.

duced by using the average concentration of an excess reagent? For rate data of the customary precision, in which initial concentrations are known to 1 percent accuracy and a precision of a few percent in the rate constant is desired, a minimum 10-fold excess usually suffices. Of course, accuracy will increase if a higher ratio is employed.[15,16]

Consider a reaction following a mixed-second-order rate equation. With a 10-fold excess of B over A, the concentration of B decreases during the run to $0.90[B]_0$ if the reaction has $1:1$ stoichiometry. To obtain the actual second-order rate constant, one divides the pseudo-first-order rate constant (which we designate k_{app} or k_{obs}) by $[B]_{av}$, the mean concentration of B in the interval of the reaction used for the rate determination. Assuming the kinetic data were collected over the entire course of the reaction,

$$k = \frac{k_{app}}{[B]_{av}} = \frac{k_{app}}{0.95[B]_0} \tag{2-40}$$

The order with respect to $[A]$, the limiting reagent, is known to be unity from the fit of the data to first-order kinetics and perhaps from the constancy of k_{app} at constant $[B]$ and varying $[A]$. But how does one establish the reaction order with respect to the concentration of the excess reagent, B? It might be done by carrying out experiments in which each successive reagent such as B is made the limiting one, all other reagents, including A, being present in large excess. In practice, this is less likely to be done than additional experiments in which A remains the limiting reagent and the concentration of each of the excess reagents is varied, in turn, over as wide a range as the experimental technique and other considerations allow.

Suppose one determines values of k_{app} over a range of $[B]_0$. If k computed according to Eq. (2-40) is constant within acceptable error limits, then that aspect of the problem is settled and a first-order dependence on $[B]$ is affirmed. On the other hand, if $k_{app}/[B]$ is not constant, then one needs to seek an alternative functional form. Depending on circumstances, that may involve one, two, and possibly as many as three rate parameters which will be optimized in the final data analysis. One might arrive at the correct functional form by trial and error. Plots of $\log k_{app}$ versus $\log[B]$ are useful guides to the formulation of plausible trial functions. (Refer to Fig. 1-1 and to the equation relating k_{app} to $[H^+]$ in that case.) The trial function is then tested by a graphical or computational procedure. The former is especially suitable if only one or two parameters are involved, because a linear relation to test the trial rate law is readily found.

The trial rate law for $CrCl^{2+}$ (Sec. 1-3) is

$$k_{app} = k_1 + \frac{k_2}{[H^+]} \tag{2-41}$$

Linearity of the plot of k_{app} versus $1/[H^+]$, as shown in Fig. 2-7, affirms the form of Eq. (2-41). The slope and intercept give the values of k_1 and k_2.

One is obligated to test data over a wide range of $[B]$ to establish the rate dependence on $[B]$. If the relation seems to be a simple one, such as the direct first-order dependence of Eq. (2-41), then a narrower range is possible than if the rate law has

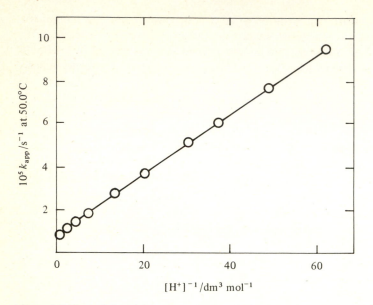

Figure 2-7 Test of the trial rate equation (2-41) for decomposition of $CrCl^{2+}$ by a plot of k_{app} (at 50.0°C) versus $1/[H^+]$. (See Ref. 2, chap. 1.)

two or more terms. Multiterm rate laws can take a number of forms; the most common two- and three-parameter expressions for k_{app} are

$$k + k'[B] \qquad\qquad k[B] + k'[B]^2$$

$$(k + k'[B])/(1 + k''[B]) \qquad k[B]/(k' + [B])$$

$$k + k'/[B] \qquad\qquad k + k'[B] + k''/[B]$$

The *clue* to which form is the correct one may come from the plot of log k_{app} versus log $[B]$; the *test* of its correctness is the fit of the data to the relation itself.

As convenient as the method is, flooding carries with it one danger. The observations are particularly susceptible to effects of a reactive impurity in the major components. Compound B may contain a trace impurity X which is highly reactive with A; and although $[X]$ may be $\ll [B]$, swamping the system such that $[B] \gg [A]$ may provide appreciable amounts of X relative to A. Hence the observation presumed to be the reaction of A and B, may, in fact, be the unsuspected reaction of X and A. The careful worker uses different and independent syntheses of samples of B or shows that the rate remains invariant as a given sample of B is repurified.

2-8 THE EXPERIMENTAL DETERMINATION OF REACTION ORDERS

The preceding section addressed the question of the order with respect to the reagent(s) in excess; the question now at hand is the order with respect to $[A]$, the limiting reagent. (In the usual laboratory situation these considerations may arise in opposite order.) Despite the emphasis placed on it by many writers, this topic does not present much difficulty if the rate consists of a single term with a whole-number order in [A]. It is generally preferable to separate the question of the order in [A] from the evaluation of k_{app}. The former should be done first. After a preliminary rate expression is established, perhaps by one of the methods described in this section, its adequacy is then checked by the quality of the fit to the integrated expression by using a least-squares or graphical or computer fit. Of course, if the fit is poor or the scatter in k is outside the acceptable range, the rate expression first deduced must be reassessed.

We shall consider three methods for determining the order with respect to [A]: (1) integrated rate laws, (2) fractional lifetime methods, and (3) the use of E, the extent of reaction.

Testing for simple first-order or second-order kinetics by a plot of ln [A] versus time and/or a plot of [A] versus [A] $\cdot t$ is done easily and quickly. Only a few data points well spaced throughout the course of reaction are needed. It is important to include points to at least 75 percent completion. Both plots will be nearly linear during the first 50 percent of reaction, and a mistaken conclusion might be drawn. To illustrate the problem, consider data presented earlier in this chapter for reactions following first-order kinetics (the hydrolysis of triphenylmethyl chloride, Table 2-1) and second-order kinetics (the disproportionation of p-toluenesulfinic acid, Table 2-2). Figure 2-8 shows the data from each reaction plotted as if a first-order equation applied to each. Although the second-order data clearly are not linear on this plot, curvature is noticeable only beyond ~50% reaction. The same point can be made when the same sets of data are plotted according to a second-order rate law. Kinetic plots of data taken over only 50% reaction do not provide a reliable determination of the reaction order.

Keep in mind, too, how small errors in the choice of the end-point reading P_∞, when an instrumental method is being used, can cause appreciable curvature late in a run. Therefore, the reaction order can be established more convincingly by the constancy of the rate constant when evaluated in experiments over a range of $[A]_0$.

In many real situations, workers are often guided in their choice of trial rate expressions by literature precedents and by likely mechanisms. That often facilitates the selection of possible rate expressions. If it is not successful, one might attempt to determine the instantaneous rate of reaction at different points throughout the run. If the rate is steadily decreasing over the course of the measurements, the values of rate determined by drawing tangents or approximated by calculation of $\Delta[A]/\Delta t$ over small increments of reaction will be quite imprecise. Nevertheless, the data may be sufficient to suggest a trial rate function, often by examination of a plot of log rate against log [A]. The trial function so selected can then be tested in its integrated form.

One circumstance in which the rate itself can be evaluated with some accuracy will be considered. If there is a product which can be determined at a very low concentration level, say by following the growth of an intense absorption peak of one product, then its initial rate (which is constant over this interval) can be accurately evaluated in the early stages. If the reaction is followed only 1 or 2 percent toward completion, the concentrations of the reactants remain effectively constant. The rate over the interval will be constant as well. A set of such determinations with varying reactant concentrations will not only determine the form of the kinetic expression but also provide pre-

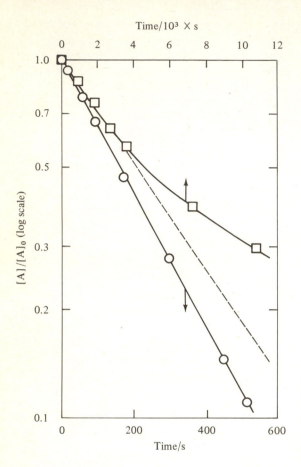

Figure 2-8 Kinetic data for first-order and for second-order reactions plotted as if both were first-order. Data for the first-order reaction are from Table 2-1, and for the second-order reaction from Table 2-2. First-order (○); second-order (□).

cise rate constants as well. This method, like the flooding technique, is susceptible to influence by reactive impurities.

The fractional time approach is based upon the concept that the time required to consume a given fraction, say half, of the limiting reagent is a characteristic of the rate equation. A comparison of successive half-times (or of any other convenient fractional time) reveals whether a reaction follows any simple integral-order rate equation. Thus the ratio of the time required for 75% reaction to that for 50% reaction is characteristic of the reaction order. Values of $t_{3/4}/t_{1/2}$ are equal to 1.5, 2, 3, and 5 for reactions of zero, first, second, and third order in a single reactant A.

Several methods are based on equations for E, the extent of reaction, defined by Eq. (2-42). The quantity is readily calculated from the recording of P against t. Regardless of the form of the kinetics, E is just the fractional progress of P from its starting to its final reading, Eq. (2-43).

$$E = 1 - \frac{[A]}{[A]_0} \qquad 0 < E < 1 \tag{2-42}$$

$$E = \frac{P_0 - P_t}{P_0 - P_\infty} \tag{2-43}$$

Several authors have shown how relations between E and time can be used to determine the order in $[A]$. We shall consider the method proposed by Wilkinson.[17] The derivation given consists of a binomial expansion of $(1 - E)^{1-n}$, dropping higher-order terms at long times. We present here an alternative method[18] for whole-number values of n (0, 1, and 2) which gives the same resulting equation,

$$\frac{t}{E} \simeq \frac{1}{k[A]_0^{n-1}} + \frac{n}{2} t \tag{2-44}$$

It is easily shown (Prob. 2-9) that Eq. (2-44) is exact for $n = 0$ and 2, and we need consider here only the first-order relation. For a first-order reaction, t/E is given by

$$\frac{t}{E} = \frac{t}{1 - [A]/[A]_0} = \frac{t}{1 - \exp(-kt)} \tag{2-45}$$

Expanding the term $\exp(-kt) = 1 - kt + (kt)^2/2 - (kt)^3/12 + \cdots$,

$$\frac{t}{E} = \frac{t}{kt - (kt)^2/2 + \cdots} = \frac{1}{k} + \frac{t}{2} - \frac{kt^2}{12} + \cdots \tag{2-46}$$

Dropping higher-order terms,

$$\frac{t}{E} \cong \frac{1}{k} + \frac{t}{2} \tag{2-47}$$

which agrees with Wilkinson's relation, Eq. (2-44), for $n = 1$.

An illustration of this method is provided by a plot of t/E versus t shown in Fig. 2-9, in which the same first- and second-order kinetic data shown earlier (Tables 2-1 and 2-2) are used. The plots are adequately linear, at least over the range where they are expected to be ($E < 0.4$ for $n = 1$), and the slope gives a value of the reaction order sufficient to distinguish one whole-number reaction order from another. (The intercept provides an estimate, but not an accurate value, of k.) The method distinguishes first- and second-order kinetics based upon data early in a run, a distinction we saw earlier was nearly impossible from the usual plots of the integrated rate law. Some caution may be called for,[18] however, in that relatively small errors in concentration can lead to a large error in E early in the run.

2-9 MULTITERM RATE EXPRESSIONS

A more complicated situation arises if the order with respect to the limiting reagent does not correspond to the simple rate law $-d[A]/dt = k[A]^n$. In such cases the integrated expression may be more cumbersome. Two cases of this type will be considered. The first is a two-term rate equation consisting of parallel first- and second-order kinetic terms. The reaction rate is given as the sum of the two rates

$$\frac{-d[A]}{dt} = k_1[A] + k_2[A]^2 \tag{2-48}$$

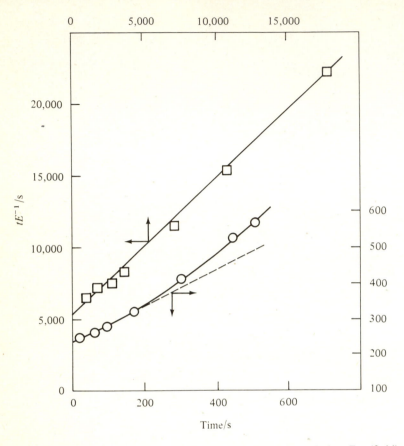

Figure 2-9 Plots of t/E versus t, according to the Wilkinson relation, Eq. (2-44), for a first-order reaction (○, Table 2-1) and a second-order reaction (□, Table 2-2). The respective slopes in the first 40% reaction are 0.50 ($n = 1$) and 0.95 ($n = 2$).

The rate expression, after integration, can be arranged in the form

$$\exp\left(k_1 t + \ln \frac{[A]}{[A]_0}\right) = \frac{k_1 + k_2[A]}{k_1 + k_2[A]_0} \tag{2-49}$$

If it is possible to conduct runs having $[A]_0$ sufficiently low, such that the second-order kinetic term contributes negligibly, then k_1 can be determined reasonably well and data from runs with higher $[A]_0$ can be treated as follows. Values on the left side of Eq. (2-49) can be calculated at each point in the run, and the linear plot of the quantity against $[A]$ affords a slope and intercept from which k_2 can be calculated.

   An example of that form was encountered during a study[19] of hydrido-bis(dimethylglyoximato)cobalt(III) phosphine:

$$2HCo(dmgH)_2PR_3 = 2Co(dmgH)_2PR_3 + H_2 \tag{2-50}$$

The kinetic data at constant $[H^+]$ followed Eq. (2-48), where A = the hydrido complex. A plot of the data for a typical run in the manner suggested, using the known value of k_1, is given in Fig. 2-10.

The second two-term rate law which we shall consider contains an autocatalytic contribution. Imagine that the mechanism of the reaction A + B = C follows two parallel pathways. The first is the direct reaction, and the second is a bimolecular reaction of B and C (the product). (In the second case, the rate-limiting step is followed by a fast reaction.)

Path 1: $$A + B \xrightarrow{k_1} C$$

Path 2: $$B + C \xrightarrow{k_2} X$$

$$X + A \xrightarrow{\text{fast}} 2C$$

The reaction rate is given by

$$\frac{-d[A]}{dt} = k_1 [A][B] + k_2 [B][C] \tag{2-51}$$

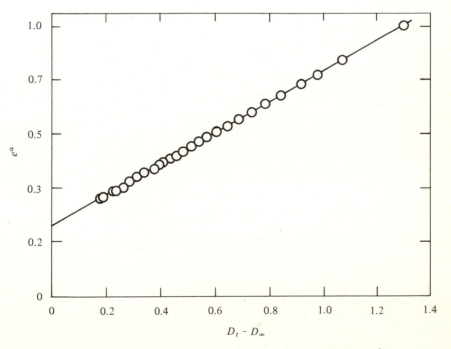

Figure 2-10 Kinetic data for reaction (2-50), with the kinetic data plotted as e^α versus $D_t - D_\infty$ according to Eq. (2-49). The symbols are D = absorbance, $\alpha = k_1 t + \ln ([A]/[A]_0)$. $D_t - D_\infty$ is directly proportional to [A]. The value of k_2 is given by $k_2 = (\text{slope})(k_1)(D_0 - D_\infty)/(\text{intercept})[A]_0$. Data from Ref. 19.

If $[C]_0 = 0$, then $[A] + [C] = [A]_0$; furthermore, if the reaction is carried out with $[B]_0 \gg [A]_0$, the equation becomes

$$\frac{-d[A]}{(k_1 - k_2)[A] + k_2[A]_0} = [B]_{av}\, dt \tag{2-52}$$

Integration gives the result

$$(k_1 - k_2)[A] + k_2[A]_0 = k_1[A]_0 \exp\{-(k_1 - k_2)[B]_{av}t\} \tag{2-53}$$

In this case the reaction does not follow pseudo-first-order kinetics, *yet the Guggenheim plot is linear* (see Prob. 2-7) and affords a value of the rate constant difference, $k_1 - k_2$. It may be possible to evaluate k_2 independently from a study of the reaction of A and C.

The above situation has been encountered during the oxidation of the tantalum cluster cation $Ta_6Br_{12}^{2+}$ (=A) to $Ta_6Br_{12}^{3+}$ (=C) by vanadium(V)(=B).[20] The autocatalytic reaction of $Ta_6Br_{12}^{3+}$ and vanadium(V) produces $Ta_6Br_{12}^{4+}$ (=X), but the latter does not accumulate in the system owing to its rapid reaction with $Ta_6Br_{12}^{2+}$.

REFERENCES

1. Margerison, D.: in C. H. Bamford and C. F. H. Tipper (eds.), "Comprehensive Chemical Kinetics," Elsevier Scientific Publishing Co., Inc., New York, vol. 1, pp. 370–386.
2. Swain, C. G., C. B. Scott, and K. H. Lohmann: *J. Am. Chem. Soc.*, 75:136 (1953).
3. Kice, J. L., and K. W. Bowers: *J. Am. Chem. Soc.*, 84:605 (1962); K. W. Bowers, Ph.D. thesis, Univ. South Carolina, 1962.
4. Miller, F. M., and M. L. Adams: *J. Am. Chem. Soc.*, 75:4599 (1953).
5. (*a*) Glasstone, S.: "Textbook of Physical Chemistry," 2d ed., D. Van Nostrand Company, New York, 1946, p. 1055; (*b*) S. Lowell, *J. Chem. Educ.*, 42:552 (1965).
6. Wiberg, K.: in E. S. Lewis (ed.), "Investigations of Rates and Mechanisms of Reactions: General Considerations and Reactions at Conventional Rates," vol. VI, part 1 of the series "Techniques of Chemistry," Wiley-Interscience, New York, 1974, pp. 750–752.
7. Guggenheim, E. A.: *Phil. Mag.*, 2:538 (1926); see also E. L. King, *J. Am. Chem. Soc.*, 74:563 (1952).
8. Kezdy, F. J., J. Kaz, and A. Bruylants: *Bull. soc. chim. Belges*, 67:687 (1958).
9. Swinbourne, E. S.: *J. Chem. Soc.*, 2371 (1960); see also P. C. Manglesdorf, *J. Appl. Phys.*, 30:443 (1958).
10. Westheimer, F. H., and H. Cohen: *J. Am. Chem. Soc.*, 60:91 (1938).
11. Hartley, H. O.: *Biometrika*, 35:32 (1948).
12. Moore, P.: *J. Chem. Soc. Faraday I*, 68:1890 (1972).
13. Espenson, J. H.: *J. Chem. Educ.*, 57:160 (1980).
14. Newton, T. W.: *J. Phys. Chem.*, 62:943 (1958).
15. Sicilio, F., and M. D. Peterson: *J. Chem. Educ.*, 38:576 (1960).
16. Corbett, J. F.: *J. Chem. Educ.*, 49:663 (1972).
17. Wilkinson, R. W.: *Chem. Ind., London*, 1395 (1961).
18. King, E. L.: private communication.
19. Chao, T.-H., and J. H. Espenson: *J. Am. Chem. Soc.*, 100:129 (1978).
20. Espenson, J. H.: *Inorg. Chem.*, 7:631 (1968).

PROBLEMS

2-1 *First-order kinetics.* A reaction following first-order kinetics was studied by determining the reactant concentration at equally spaced time intervals. Each successive pair of concentrations—A_0 and A_1, A_1 and A_2, etc.—was used to calculate k, and the average of the individual k's was taken as the best value. Show this to be equivalent to using the first value and the last, ignoring the intermediate values.

2-2 *Guggenheim's and Swinbourne's methods.* Nann, Powell, and Hall [*J. Chem. Soc. B*, 1683 (1971)] studied the reaction of 7,7,8,8-tetracyanoquinodimethane (TCNQ) with triphenylphosphine in aqueous acetonitrile in the presence of HCl. The reaction was followed by the disappearance of TCNQ at 394 nm in the presence of a large excess of phosphine. Values of absorbance at intervals of 5.00 s, starting ca. 15 s after the start of the run, were 0.560, 0.530, 0.500, 0.473, 0.447, 0.421, 0.399, 0.376, 0.356, 0.337, 0.320, 0.302, 0.288, 0.272, 0.259, 0.246, 0.233, 0.221, 0.210, 0.200, 0.191.

(*a*) Use both Guggenheim's and Swinbourne's methods to compute the pseudo-first-order rate constant.

(*b*) For each method compute the final absorbance reading and compare with the experimental value of 0.024.

2-3 *Second-order kinetics.* A reaction which is second-order with respect to a single component was 75 percent complete in 92 min when the initial concentration was 0.240 mol dm^{-3}. How long would be required to achieve a concentration of 0.043 mol dm^{-3} in a run with an initial concentration of 0.146 M?

2-4 *Equivalent concentrations.* The reaction $2A + B = C + D$ was studied under conditions $[A]_0 = 2[B]_0$; a plot of $[A]$ versus $[A] \cdot t$ was linear.

What rate expression(s) does this suggest, and what experiments could be designed to provide further detail?

2-5 *Second-order kinetic data.* Derive Eq. (2-35), and apply it to the second-order reaction $Np^{3+} + Fe^{3+} = Np^{4+} + Fe^{2+}$ [*J. Phys. Chem.*, 71:3768 (1967)]. Conditions: 298 K, $[H^+] = 0.400$ mol/dm^3, ionic strength = 2.00 mol/dm^3. Initial concentrations: Np^{3+} 1.58×10^{-4}; Fe^{3+} 2.24×10^{-4} mol/dm^3.

Time/s	0	2.5	3.0	4.0	5.0	7.0	10.0	15.0	20.0
Absorbance 723 nm, b = 5 cm	0.100	0.228	0.242	0.261	0.277	0.300	0.316	0.332	0.341

The final reading is 0.351, at completion.

2-6 *Instrumental methods in a pseudo-zero-order reaction.* The bromination of acetone is believed to follow this mechanism:

$$(CH_3)_2CO \xrightarrow{k_1} CH_3\overset{\displaystyle OH}{\underset{\displaystyle |}{C}}{=}CH_2 \xrightarrow[\text{Very fast}]{Br_2} CH_3\overset{\displaystyle O}{\underset{\displaystyle ||}{C}}CH_2Br + H^+ + Br^-$$

When studied under the conditions $[(CH_3)_2CO] \gg [Br_2]$, the reaction follows pseudo-zero-order kinetics, $-d[Br_2]/dt = k_1[(CH_3)_2CO] = $ constant.

The reaction rate can be evaluated spectrophotometrically. In one experiment at 20.3°C in 0.40 N H_2SO_4, having $[(CH_3)_2CO] = 0.645$ mol/dm^3 and $[Br_2] = 0.0193$ mol/dm^3, the following data were obtained. Derive the integrated rate expression applicable to this situation and use it to calculate k_1.

Time/s	0	600	1200	3600	4800	6000	9000
Absorbance	0.201	0.257	0.313	0.558	0.665	0.683	0.683

2-7 *Autocatalytic kinetics by Guggenheim's method.* The kinetics of oxidation of $Ta_6Br_{12}^{2+}$ by V(V) have been given in the text, Eq. (2-51).

(a) Show that in experiments with $[V(V)]_0 \gg [M^{2+}]_0$, the integrated equation is

$$\ln \frac{(k_{23} - k_{34})[M^{2+}]_t + \alpha}{(k_{23} - k_{34})[M^{2+}]_0 + \alpha} = -(k_{23} - k_{34})[V(V)]t$$

where $\alpha = k_{34}[M^{2+}]_0$ and $M^{2+} = Ta_6Br_{12}^{2+}$

Note:

$$\int \frac{dx}{ax + b} = a^{-1} \ln(ax + b) + I$$

(b) Note that the pseudo-first-order relation, $\ln[M^{2+}]$ versus time, will not be linear. Apply the Guggenheim method to this system and show that $\ln([M^{2+}]_t - [M^{2+}]_{t+\tau})$ versus time will be linear. What rate constant or composite is represented by the slope of this plot?

(c) Carry out the indicated computations on this experiment. Conditions: $25°C$, $[H^+] = 0.500$ M; $[M^{2+}] = 1.1 \times 10^{-5}$; $[V(V)] = 4.00 \times 10^{-4}$.

Time/s	0	10	20	30	40	50
$10^5 [M^{2+}]/M$	1.10	0.721	0.451	0.270	0.136	0.041

2-8 *Flooding; concentration dependence.* The nucleophilic reaction of phenyl α-disulfone with hydrazine according to the reaction

$$PhSO_2SO_2Ph + N_2H_4 = PhSO_2NHNH_2 + PhSO_2H$$

was studied by Kice and Legan [*J. Am. Chem. Soc.*, **95**:3912 (1973)]. In runs with $[sulfone]_0 = 3 \times 10^{-5}$ M the reaction follows pseudo-first-order kinetics. The values of k_{app} vary with $[N_2H_4]$ as follows. Formulate the rate expression and evaluate the constant(s) therein.

$[N_2H_4]/mol\,dm^{-3}$	5×10^{-3}	1.0×10^{-2}	1.6×10^{-2}	2.0×10^{-2}	3.0×10^{-2}	4.0×10^{-2}
k_{app}/s^{-1}	0.085	0.176	0.30	0.41	0.67	0.95

2-9 *Wilkinson's Method.* Show that Eq. (2-44) is an exact relation for zero- and second-order reactions.

2-10 *Second-order kinetics with P_∞ unknown.* The following data are the absorbance readings at successive 10.0-s intervals for a reaction following the second-order rate law $-d[A]/dt = k[A]^2$, with $[A]_0 = 4.6 \times 10^{-3}$ mol dm^{-3}.

Estimate the rate constant and the end-point reading: 0.920, 0.643, 0.526, 0.462, 0.421, 0.393, 0.372, 0.353, 0.344, 0.334, 0.326, 0.319, 0.313, 0.308.

2-11 *Rate law.* The second-order rate constant for the reaction $V^{3+} + Cr^{2+} = V^{2+} + Cr^{3+}$ was evaluated as a function of $[H^+]$. From the data given below, at $25.0°C$ and constant ionic strength of 0.500 M, formulate an equation giving the dependence of k_{app} upon $[H^+]$ and evaluate the parameter(s) in the equation. Compare your answer with the data given in Prob. 1-2.

$[H^+]$/mol dm^{-3}	0.027	0.040	0.115	0.155	0.192	0.250	0.269	0.365	0.500
k/dm^3 mol^{-1} s^{-1}	5.00	4.32	2.66	2.29	2.03	1.78	1.63	1.25	1.04

2-12 *Initial rate method.* Consider a first-order reaction followed by the method of initial rates. Show that the rate constant is given by

$$k = \frac{\Delta P/\Delta t}{P_\infty}$$

where $\Delta P/\Delta t$ is the slope of the initial linear portion of the recording of P against t.

THREE

KINETICS OF COMPLEX REACTIONS—REVERSIBLE AND CONCURRENT REACTIONS

Largely, until now, we have considered only reactions whose rate law consists of a single term with an integral reaction order. Many real chemical systems are more complex than that, however, and they are characterized by rate expressions consisting of two or more terms. Three general categories can be recognized: (1) reversible or opposing reactions which attain a finite equilibrium, (2) parallel reactions producing either the same or different products from a single reactant or from a mixture of reactants, and (3) consecutive, multistep reactions in which intermediates play a key role. In this chapter we will consider the first two of those situations, which are unrelated to one another. The third is taken up in Chap. 4.

3-1 REVERSIBLE FIRST-ORDER REACTIONS

Consider the reaction

$$A \underset{k_{-1}}{\overset{k_1}{\rightleftharpoons}} B \tag{3-1}$$

in which the net rate of disappearance of A is given by

$$\frac{-d[A]}{dt} = k_1[A] - k_{-1}[B] \tag{3-2}$$

To solve Eq. (3-2), two relations are used, one from the stoichiometry [Eq. (3-3)] and the other from the condition $-d[A]/dt = 0$ at equilibrium [Eqs. (3-4) and (3-5)].

$$[A]_0 + [B]_0 = [A]_\infty + [B]_\infty = [A] + [B] \qquad (3\text{-}3)$$

$$k_1 [A]_\infty = k_{-1} [B]_\infty \qquad (3\text{-}4)$$

$$\frac{[B]_\infty}{[A]_\infty} = K = \frac{k_1}{k_{-1}} \qquad (3\text{-}5)$$

where the notations $[\ ]_\infty$ and $[\ ]_{eq}$ represent the same value. Substitution of Eqs. (3-3) and (3-5) into the rate law of Eq. (3-2), and rearrangement, affords the relation

$$\frac{-d[A]}{dt} = (k_1 + k_{-1})([A] - [A]_\infty)$$

Integration between the usual limits gives

$$\ln \frac{[A] - [A]_\infty}{[A]_0 - [A]_\infty} = -(k_1 + k_{-1})t \qquad (3\text{-}6)$$

Therefore, a plot of $\ln([A] - [A]_\infty)$ versus time for such a case will be linear. The plot gives, as the negative of the slope, the *sum* $k_1 + k_{-1}$. The sum can be resolved into the individual forward and reverse rate constants by use of the equilibrium constant, which gives the ratio $K = k_1/k_{-1}$. The result is an interesting one, and one of its implications is at first surprising. The apparent rate constant is $k_1 + k_{-1}$, which is *larger* than the forward rate constant. On the other hand, inclusion of the reverse reaction leads to a net reaction rate which is *smaller* than the forward rate alone, since the net rate consists of a difference of two terms as shown in Eq. (3-2). That illustrates the necessity of making a careful distinction between a reaction rate and a rate constant.

We can amplify this further by using a hypothetical reaction which fits the pattern. Consider a reaction for which $k_1 = 2.60 \times 10^{-2}$ s^{-1} and $k_{-1} = 4.10 \times 10^{-2}$ s^{-1}. Figure 3-1 depicts the concentration of reactant A as a function of time. Now consider this same reaction run under conditions in which there has been added some reagent which very rapidly consumes B and thus prevents its return to A. In that case only the forward rate of reaction contributes to $-d[A]/dt$, and the resulting curve of $[A]$ versus time, also shown in the figure, is controlled only by the value of k_1. The initial rate of reaction is the same in the two situations; in the first instance the return of B to A can be seen to lower the net rate more than when the reverse step is missing. On the other hand, since the reaction which approaches the finite equilibrium has much less A to convert to B in reaching equilibrium, it is not entirely surprising that the half-time to reach equilibrium is shorter than that to go to completion.

The plot of $\ln\{([A] - [A]_\infty)/([A]_0 - [A]_\infty)\}$ against time will be linear and will have an apparent rate constant of $k_1 + k_{-1}$ irrespective of whether the reaction began with pure A or pure B or a mixture of A and B or whether concentrations were such that the particular run proceeded in one direction or the other. If one were to plot merely log $[A]$ versus time, as if the reaction were irreversible, the plot would be curved and the curvature would become more pronounced for a larger k_{-1}/k_1 ratio.

Consider the situation which develops when a reaction such as the preceding one is followed by using an instrumental technique. Using the same notation as in Sec.

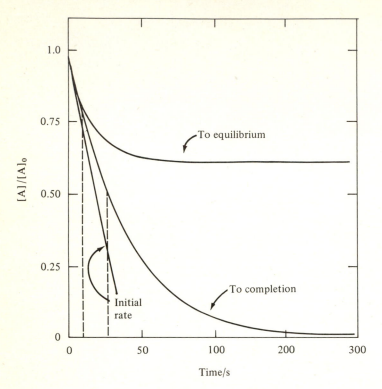

Figure 3-1 Illustrating the time dependence of [A] for a hypothetical, reversible, first-order reaction, Eq. (3-1) with $k_1 = 2.60 \times 10^{-2}$ s^{-1} and $k_{-1} = 4.10 \times 10^{-2}$ s^{-1}. Shown are [A] when the reaction proceeds to equilibrium and also when only the forward rate contributes, B being drawn off in a rapid reaction, forcing the reaction to completion. The initial reaction rate, the same in both situations, is also shown. The figure also indicates the half-time to reach completion, ln $2/k_1 = $ 26.7 s and to progress one-half the way toward equilibrium, ln $2/(k_1 + k_{-1}) = 10.3$ s.

2-4, the values of the property P are

$$P_0 = p_A[A]_0 + p_B[B]_0 + C$$

$$= (p_A - p_B)[A]_0 + p_B(1 + K)[A]_\infty + C$$

$$P_t = (p_A - p_B)[A]_t + p_B(1 + K)[A]_\infty + C$$

$$P_\infty = (p_A + Kp_B)[A]_\infty + C$$

Substitution into Eq. (3-6), and simplification, affords the equation

$$\ln \frac{[A] - [A]_\infty}{[A]_0 - [A]_\infty} = \ln \frac{P_t - P_\infty}{P_0 - P_\infty} = -(k_1 + k_{-1})t \qquad (3\text{-}7)$$

This equation suggests that the correct plot is log $(P_t - P_\infty)$ versus time or, if the value of P increases with time, log $(P_\infty - P_t)$ versus time. The slope will give an apparent rate constant; again it is the sum $k_1 + k_{-1}$, as when concentration values were used directly. Proper application of Eq. (3-7) requires an accurate value of the "end-point reading"

P_∞. As used here, this refers to the actual value at equilibrium and not to the value of P which would correspond to 100 percent completion. Were the latter used, curved kinetic plots would result, and at best an approximate solution would be obtained.

3-2 OPPOSING REACTIONS OF HIGHER ORDER

Cases such as given in Eqs. (3-8) and (3-9)

$$A + B \underset{-1}{\overset{1}{\rightleftharpoons}} C \tag{3-8}$$

$$A + B \underset{-1}{\overset{1}{\rightleftharpoons}} C + D \tag{3-9}$$

become increasingly complex, and complete solutions are shown in Table 3-1. The equations are manageable under some circumstances, particularly if the equilibrium constant is known. In that case, the relative values of k_1 and k_{-1} are fixed by the relation $K = k_1/k_{-1}$. Although such systems can be solved, the results are cumbersome. In practice one often chooses to simplify the kinetic treatment by selection of concentration conditions. If all concentrations but one on each side of the reaction are high, the kinetic equations reduce to a set of opposing pseudo-first-order reactions. For reaction (3-8), for example, the net rate is

$$\frac{d[A]}{dt} = -k_1[A][B] + k_{-1}[C] \tag{3-10}$$

If one does experiments with $[B] \gg [A]$, then (3-10) reduces in effect to a case of opposing first-order reactions. A plot of $\log([A] - [A]_\infty)$ versus time is linear, and the apparent rate constant is given by

$$k_{app} = \frac{-d\ln([A] - [A]_\infty)}{dt} = k_1[B] + k_{-1} \tag{3-11}$$

In this case, when $[B]$ is varied in different runs, k_1 and k_{-1} are obtained from a plot of k_{app} versus $[B]$. The slope and intercept afford values of the individual rate constants k_1 and k_{-1}, and their quotient gives the equilibrium constant.

In practice, a reliable value of K might be available from accurate static (equilibrium) measurements. In that case, the quotient k_1/k_{-1} can be compared with the known value. At that, it is perhaps preferable to introduce the reliable K value into the kinetic computation and thereby reduce the number of parameters needed to describe the kinetics. Equation (3-11) can be rearranged to

$$k_{app} = k_{-1}\left(\frac{k_1}{k_{-1}}[B] + 1\right) = k_{-1}(K[B] + 1) \tag{3-12}$$

$$\frac{k_{app}}{K[B] + 1} = k_{-1} \tag{3-13}$$

Table 3-1 Complete equilibrium and kinetic expressions for selected reversible reactions

System 1

$$A + B \underset{k_r}{\overset{k_f}{\rightleftharpoons}} C \qquad K = \frac{[C]_\infty}{[A]_\infty[B]_\infty} = \frac{k_f}{k_r}$$

$[A]_0 = [B]_0$
$[C]_0 = 0$

$$[A]_\infty = \frac{1}{2K}\left(\sqrt{1+4K[A]_0} - 1\right)$$

$$k_f = \frac{K}{t(2K[A]_\infty + 1)} \ln \frac{([A] + [A]_\infty + 1/K)([A]_0 - [A]_\infty)}{([A]_0 + [A]_\infty + 1/K)([A] - [A]_\infty)}$$

$[B]_0 > [A]_0$
$[C]_0 = 0$
$\Delta = [B]_0 - [A]_0$

$$[A]_\infty = \frac{1}{2K}\left\{\sqrt{(1+K\Delta)^2 + 4K[A]_0} - (1+K\Delta)\right\}$$

$$k_f = \frac{K}{t(2K[A]_\infty + 1 + K\Delta)} \ln \frac{([A] + [A]_\infty + 1/K + \Delta)([A]_0 - [A]_\infty)}{([A]_0 + [A]_\infty + 1/K + \Delta)([A] - [A]_\infty)}$$

System 2

$$A + B \underset{k_r}{\overset{k_f}{\rightleftharpoons}} C + D \qquad K = \frac{[C]_\infty[D]_\infty}{[A]_\infty[B]_\infty} = \frac{k_f}{k_r}$$

$[A]_0 = [B]_0$
$[C]_0 = [D]_0 = 0$

$$[A]_\infty = [A]_0 \left(\frac{1-\sqrt{K}}{1-K} \right)$$

$$k_f = \frac{\sqrt{K}}{t\,2[A]_0} \ln \frac{([A]_0 - [A]_\infty)\{[A] + [A]_\infty + 2[A]_0/(K-1)\}}{([A] - [A]_\infty)\{[A]_0 + [A]_\infty + 2[A]_0/(K-1)\}}$$

$[B]_0 > [A]_0$
$[C]_0 = [D]_0 = 0$
$\Delta = [B]_0 - [A]_0$

$$[A]_\infty = \frac{1}{2(K-1)}\left\{\sqrt{(\Delta K + 2[A]_0)^2 + 4[A]_0^2(K-1)} - (\Delta K + 2[A]_0)\right\}$$

$$k_f = \frac{K}{t\,2(K-1)[A]_\infty + \Delta K + 2[A]_0} \ln \frac{([A]_0 - [A]_\infty)\{[A] + [A]_\infty + (\Delta K + 2[A]_0)/(K-1)\}}{([A] - [A]_\infty)\{[A]_0 + [A]_\infty + (\Delta K + 2[A]_0)/(K-1)\}}$$

or, with similar substitutions, to the form

$$\frac{k_{app}}{[B] + K^{-1}} = k_1 \qquad (3\text{-}14)$$

Hence either Eq. (3-13) or Eq. (3-14) can be used together with the independently known value of K to compute one rate constant. The other is then calculated from the relation $K = k_1/k_{-1}$.

As an example, consider published results[1] on the reaction

$$Co(edta)^{2-} + Fe(CN)_6^{3-} \underset{k_{-1}}{\overset{k_1}{\rightleftharpoons}} [(edta)Co\text{-}NC\text{-}Fe(CN)_5]^{5-} \qquad (3\text{-}15)$$

for which the equilibrium constant is 1.40×10^3 M^{-1} (5.75°C, pH 6.00). The reaction was studied with a large excess of the cobalt reactant, and the rate of approach to equilibrium was found to follow a first-order expression. The values of k_{app} increased linearly with $[Co(edta)^{2-}]$, as shown in Fig. 3-2. That is in accord with the expressions

$$k_{app} = k_1[Co(edta)^{2-}] + k_{-1} \qquad (3\text{-}16)$$

$$\frac{-d[Fe(CN)_6^{3-}]}{dt} = k_1[Co(edta)^{2-}][Fe(CN)_6^{3-}] - k_{-1}[(edta)Co\text{-}NC\text{-}Fe(CN)_5^{5-}] \quad (3\text{-}17)$$

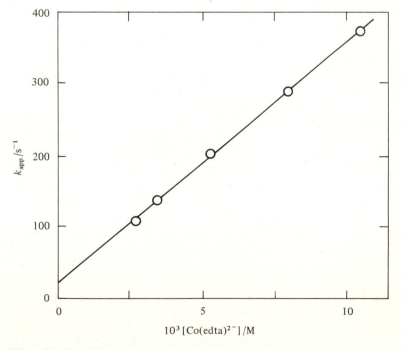

Figure 3-2 Plot of the apparent first-order rate constant for the approach to equilibrium in reaction 3-15 as a function of $[Co(edta)^{2-}]$, the reagent present in large excess. The slope and intercept afford $k_1 = 3.36 \times 10^4$ M^{-1} s^{-1} and $k_{-1} = 24$ s^{-1} in accord with Eq. (3-16). Data are from Ref. 1.

Table 3-2 Resolution of the forward and reverse rate constants for reaction (3-15)^a using the value of $K = 1.4 \times 10^3$ M^{-1}

10^3 [Co(edta)$^{2-}$]/M	k_{app}/s^{-1} at 5.75°C	$10^{-4} k_1$/M^{-1} s^{-1} from Eq. (3-14)	k_{-1}/s^{-1} from Eq. (3-13)
2.63	109	3.26	23.3
3.42	137	3.31	23.7
5.26	203	3.40	24.3
7.89	286	3.32	23.7
10.5	373	3.33	23.8
		Av: 3.36 ± 0.04	23.8 ± 0.4

^{a}Data from Ref. 1.

The intercept of the plot of k_{app} versus [Co(edta)$^{2-}$], Fig. 3-2, is $k_{-1} = 24$ s^{-1}, and from the slope $k_1 = 3.36 \times 10^4$ M^{-1} s^{-1}. An alternative method of treating the kinetic data is to compute values of k_1 and k_{-1} for each experiment by using the relations in Eqs. (3-13) and (3-14) with the independently known value of the equilibrium constant. The results are summarized in Table 3-2.

3-3 "CONCENTRATION-JUMP" METHODS FOR REVERSIBLE EQUILIBRIA

"Concentration-jump" methods refers to a method of experimental design often considered in conjunction with special methods for the study of very rapid reactions. The method is equally applicable to all reversible systems, however, and it can properly be considered at this time. It is easy to imagine a reversible reaction which cannot readily be studied under concentration conditions in which the system reduces to two opposing first-order reactions. Any number of situations might lead to this: The rate may be too high for the methods available at concentrations high enough for pseudo-first-order conditions; undesired, competing reactions may set in at those conditions; the detection method may be insensitive with one or more reagents added in large excess; the nature of the reaction medium may be too drastically altered at the high concentrations; etc. On the other hand, the complete integrated kinetic equation is formidable (see Table 3-1) and quite possibly not of a form that is easy to use.

Consider what happens when the concentration of one or more reagents is suddenly changed. The reaction must readjust itself to a new set of equilibrium concentrations consistent with the value of the equilibrium constant, and the rate at which that happens is governed by the rates of the opposing reactions. As shown later, if the perturbation is not too large, two very substantial simplifications will occur. (1) The rate of reequilibration will follow an apparent pseudo-first-order equation, regardless of the form of the rate law for the reaction, and that will permit ready analysis of the data from each run. (2) The value of the rate constant so computed will always bear a simple relation to the new equilibrium concentrations, which will permit affirmation of the form of the rate equation and evaluation of forward and reverse rate constants.

The concentration jump can be attained either by addition of one component or by dilution of an originally equilibrated solution with solvent. The latter method is often the more useful, but it is restricted to reactions having a different number of solute species on either side of the equation. With neither method must the extent of the perturbation be too large. That is not too serious a limitation, however, as explained in the subsequent development.

To illustrate the applicable equations, consider that the reversible system in Eq. (3-8) is perturbed by sudden dilution with an equal volume of solvent.

$$A + B \underset{-1}{\overset{1}{\rightleftharpoons}} C \tag{3-8}$$

If the original equilibrium concentrations before dilution are represented by $[A]_{eq\ 1}$, etc., then at the instant of dilution the concentration of each component is just half that value, $[A]_{eq\ 1}/2$, etc. The reaction shifts to the left to reestablish a new set of equilibrium concentrations (designated $[A]_{eq\ 2}$, etc.) whose values satisfy the expression

$$K = \frac{k_1}{k_{-1}} = \frac{[C]_{eq\ 2}}{[A]_{eq\ 2}[B]_{eq\ 2}} \tag{3-18}$$

If the quantity δ is used to represent the increment by which each concentration differs from its new equilibrium value at any given time, the concentrations of the components during the course of the reequilibration are then

$$[A] = [A]_{eq\ 2} - \delta$$

$$[B] = [B]_{eq\ 2} - \delta$$

$$[C] = [C]_{eq\ 2} + \delta$$

Substitution of those relations into the rate law for the reaction gives the following equation:

$$\frac{-d[A]}{dt} = k_1[A][B] - k_{-1}[C] = \frac{d\delta}{dt}$$

$$\frac{d\delta}{dt} = k_1([A]_{eq\ 2} - \delta)([B]_{eq\ 2} - \delta) - k_{-1}([C]_{eq\ 2} + \delta)$$

Expansion of that and combination of like terms gives an expression which can be further simplified by use of the equilibrium condition, Eq. (3-18), giving

$$\frac{-d\delta}{dt} = \{k_1([A]_{eq\ 2} + [B]_{eq\ 2}) + k_{-1}\}\delta + k_1\delta^2$$

Provided the perturbation is sufficiently small that the last term in δ^2 can be neglected in comparison with the others, the equation becomes

$$\frac{-d\ln\delta}{dt} = k_{obs} = k_{-1} + k_1([A]_{eq\ 2} + [B]_{eq\ 2}) \tag{3-19}$$

Equation (3-19) illustrates the two points made above: reequilibration follows a simple first-order kinetic expression, and k_{obs} is a simple function of concentrations.

In this example, a plot of k_{obs} against the summation of the equilibrium concentrations of A and B is linear, with slope and intercept giving values of k_1 and k_{-1}.

Expressions for other reversible reactions are derived equally easily. In this text, as in others, a general treatment is given in the course of discussing perturbation methods for the study of fast reactions near equilibrium. As developed in Sec. 10-4, an elementary reaction of the general form

$$a\text{A} + b\text{B} \underset{r}{\overset{f}{\rightleftharpoons}} d\text{D} + e\text{E}$$

which follows the rate law

$$\frac{-d[\text{A}]}{a\,dt} = k_f[\text{A}]^a[\text{B}]^b - k_r[\text{D}]^d[\text{E}]^e \tag{3-20}$$

approaches equilibrium after a small perturbation following a first-order equation with a rate constant given by:[2]

$$k_{obs} = k_f[\text{A}]^a_{eq}[\text{B}]^b_{eq}\left(\frac{a^2}{[\text{A}]_{eq}} + \frac{b^2}{[\text{B}]_{eq}}\right) + k_r[\text{D}]^d_{eq}[\text{E}]^e_{eq}\left(\frac{d^2}{[\text{D}]_{eq}} + \frac{e^2}{[\text{E}]_{eq}}\right) \tag{3-21}$$

Consider application of this method to a study of the monomer-dimer equilibrium in chromate(VI) solutions, where the principal equilibrium is

$$2\text{HCrO}_4^- \underset{k_m}{\overset{k_d}{\rightleftharpoons}} \text{Cr}_2\text{O}_7^{2-} + \text{H}_2\text{O} \tag{3-22}$$

Provided that this also represents the elementary reaction, application of Eq. (3-21) with $a = 2$, $b = 0$, $d = 1$, $e = 0$, yields

$$k_{obs} = 4k_d[\text{HCrO}_4^-] + k_m \tag{3-23}$$

The indicated plot of k_{obs} against $[\text{HCrO}_4^-]$ was linear and gave $k_d = 1.8 \pm 0.3$ M^{-1} s^{-1} and $k_m = 2.7 \pm 0.4 \times 10^{-2}$ s^{-1}. The quotient of the rate constants is $K = k_d/k_m = 67 \pm 21$ M^{-1} in reasonable agreement with the accepted value $K = 50$ M^{-1} under those conditions.[3]

The very simplification that makes this method feasible, neglect of higher-order terms in δ^2, etc., leads to one disadvantage. The relaxation to the new equilibrium will be accompanied by small changes in concentration (or property proportional to concentration), changes that are superimposed on the background of the concentrations already present. Because the measured changes will be small, the amplification needed to realize a measurable signal may lead to a much higher noise level than customary. In some cases, if one of the species undergoes a protonation equilibrium, it may be desirable to add an indicator and use its rapid response and high molar absorptivity to aid in monitoring the changes.

3-4 EXCHANGE REACTIONS

A special case of opposing reactions is a situation in which no net chemical change occurs yet a reaction is occurring nonetheless. Such reactions are termed *exchange reactions*. They constitute an important field of study; for they consist of reactions having $\Delta G° = 0$. They are processes for which a theoretical interpretation of the rate constant might be attained most easily. The absence of a net driving force and the chemical identity of reactants and products suggest that the intrinsic activation barrier might be more easily formulated from first principles. The absence of net (observable) chemical change requires that the exchange process be followed through a labeling technique, most often by isotopic labeling. Examples consist of solvent exchange, ligand exchange, and electron exchange, such as the following:

$$(CH_3)_2CO + H_2^*O = (CH_3)_2C^*O + H_2O$$

$$RBr + {}^*Br^- = R^*Br + Br^-$$

$$Cr(H_2O)_6{}^{3+} + {}^*Cr(H_2O)_6{}^{2+} = Cr(H_2O)_6{}^{2+} + {}^*Cr(H_2O)_6{}^{3+}$$

The isotopic tracer may consist of a stable isotope such as ^{18}O or ^{2}H, whose change of concentration in one component or the other may be determined mass spectrometrically, or of a radioactive isotope, whose level of incorporation or depletion in a particular component may be determined by counting techniques. It should be emphasized that the exchange processes represented by these reactions occur whether or not an isotopic tracer has been added; the latter simply provides a means of following their occurrence.

The mathematics of exchange reactions, first formulated by McKay[4] in a relation which bears his name, is essentially that of opposing reactions with an equilibrium constant of unity (see Prob. 3-14). We consider the general exchange process shown in Eq. (3-24) and define three quantities useful in expressing the relations in Eq. (3-25).

$$AX + X^* = AX^* + X \qquad (3\text{-}24)$$

$$C_A = [AX^*] + [AX] \qquad (3\text{-}25a)$$

$$C_X = [X^*] + [X] \qquad (3\text{-}25b)$$

$$C^* = [AX^*] + [X^*] \qquad (3\text{-}25c)$$

At isotopic equilibrium, the tracer will be distributed statistically between the two forms in proportion to their concentrations, i.e.,

$$[AX^*]_\infty = C^* \frac{C_A}{C_A + C_X} \qquad (3\text{-}26a)$$

$$[X^*]_\infty = C^* \frac{C_X}{C_A + C_X} \qquad (3\text{-}26b)$$

Three assumptions are made at this point, although with appropriate complications each could be removed.

1. There is no isotope effect in reaction, which is equivalent to $K = 1.000$, or $\Delta G° = 0$.
2. There is no appreciable decay of X^* during the time of the experiment.
3. There is a single exchange reaction; i.e., there are not two inequivalent X's in AX.

Three derivations of the kinetic equations will be made: (1) assuming that exchange occurs by dissociation of AX, (2) assuming that exchange occurs by attack of

X on AX, (3) making no assumption of an exchange mechanism. The point of all this will be to show that *the kinetic equation takes the same form regardless of exchange mechanism.*

Assume first that the mechanism is given by

$$AX \underset{k_{-1}}{\overset{k_1}{\rightleftharpoons}} A + X \qquad AX^* \underset{k_{-1}}{\overset{k_1}{\rightleftharpoons}} A + X^* \tag{3-27}$$

The rate of incorporation of tagged X into AX is given by

$$\frac{d[AX^*]}{dt} = -k_1[AX^*] + k_{-1}[A][X^*] \tag{3-28}$$

Since the chemical equilibrium in (3-27) lies far to the left, we can write

$$[A] = \frac{k_1 C_A}{k_{-1} C_X} \tag{3-29}$$

Substitution and rearrangement give the result

$$\frac{d[AX^*]}{dt} = \frac{k_1(C_A + C_X)}{C_X} \{[AX^*]_\infty - [AX^*]\} \tag{3-30}$$

which upon integration between the usual limits becomes

$$\ln\left(1 - \frac{[AX^*]}{[AX^*]_\infty}\right) = -\frac{k_1(C_A + C_X)}{C_X} t \tag{3-31}$$

A second derivation assumes a bimolecular displacement mechanism,

$$AX + X^* \underset{k}{\overset{k}{\rightleftharpoons}} AX^* + X \tag{3-32}$$

to which the following relations apply

$$\frac{d[AX^*]}{dt} = k[AX][X^*] - k[AX^*][X] \tag{3-33}$$

$$\frac{d[AX^*]}{dt} = k(C_A + C_X)([AX^*]_\infty - [AX^*]) \tag{3-34}$$

Integration as before gives

$$\ln\left(1 - \frac{[AX^*]}{[AX^*]_\infty}\right) = -k(C_A + C_X)t \tag{3-35}$$

Finally, a general derivation is presented. Whatever the form of the actual exchange rate in a given run (a quantity designated as R_{ex}), only some of the exchange events lead to a change in the level of tagged groups incorporated into AX.

The net rate of appearance of tag in the species AX* is given by the total reaction rate R_{ex} times the fraction of such events that lead to transfer of X*. (That is, the events AX + X and AX* + X*, although a part of R, do not add or remove X* from AX.)

The rate of X^* exchange then becomes

$$\frac{d[AX^*]}{dt} = R_{ex} \frac{[AX]}{C_A} \frac{[X^*]}{C_X} - R_{ex} \frac{[AX^*]}{C_A} \frac{[X]}{C_X} \tag{3-36}$$

where R_{ex} represents the total reaction *rate* and its concentration dependences are unspecified. Substitution and rearrangement give the equation

$$\frac{d[AX^*]}{dt} = \frac{R_{ex}(C_A + C_X)}{C_A C_X} ([AX^*]_\infty - [AX^*]) \tag{3-37}$$

which integrates to give

$$\ln \left(1 - \frac{[AX^*]}{[AX^*]_\infty}\right) = -R_{ex} \frac{C_A + C_X}{C_A C_X} t \tag{3-38}$$

which is known as the McKay equation. It is often written in terms of F, the fraction of exchange occurring, and becomes[†]

$$\ln (1 - F) = -R_{ex} \frac{C_A + C_X}{C_A C_X} t \tag{3-39}$$

Note the following features of the result: the approach to exchange equilibrium follows *first-order kinetics, irrespective of the rate and rate law* of the exchange process. To illustrate that point, compare the result from derivation 1 as embodied in Eq. (3-31) if we substitute the expression $R_{ex} = k_1 C_A$ into the general solution, Eq. (3-39). Likewise, substitution of the rate assumed in derivation 2, $R_{ex} = k_1 C_A C_X$ into Eq. (3-39), leads to the same result as given in Eq. (3-35).

Probably the most convenient working procedure for exchange data is to plot $1 - F$ on semilog paper against t and read off $t_{1/2}$, the time at which the fraction of exchange is exactly $1/2$. R_{ex} is then calculated from

$$R_{ex} = \frac{C_A C_X}{C_A + C_X} \cdot \frac{\ln 2}{t_{1/2}} \tag{3-40}$$

This process is repeated for different values of C_A and C_X, and the dependence of R_{ex} upon C_A and C_X is thereby deduced. Note that if C_A or C_X is varied or both are varied, the variation in $t_{1/2}$ does not directly reflect the variation in R_{ex}, owing to the factor $(C_A C_X)/(C_A + C_X)$. If the expression for R_{ex} is $k[AX][X]$, then the value of $t_{1/2}$ is constant in experiments in which $C_A + C_X$ is constant.

As an example of an exchange reaction, consider the exchange of iodine between benzyl iodide and free iodide ions. The exchange rate was evaluated[5] by mixing the former with solutions of radiolabeled potassium iodide containing iodine-131. Samples

[†]In the general case of n equivalent exchanging groups

$$AX_n + nX^* = AX_n^* + nX$$

the expression is

$$\ln (1 - F) = -R_{ex} \frac{C_A + C_X}{n C_A C_X} t \tag{3-39a}$$

Table 3-3 Kinetic data[a] for the isotopic exchange of benzyl iodide and potassium iodide

Part A. Data from one run with $[C_6H_5CH_2I]$ = 0.01020 M and [KI] = 7.36×10^{-3} M

Time/s	$1 - F$	
0	1.00	$t_{1/2}$ = 1410 s[b]
900	0.64	
1500	0.48	R_{ex} = 2.10×10^{-6} M s^{-1}[c]
2200	0.34	

Part B. Summary of data from other runs

$10^3 [C_6H_5CH_2I]$/M	10^3 [KI]/M	$t_{1/2}$/s	$10^6 R_{ex}$/M s^{-1}	k/M^{-1} s^{-1}[d]
3.11	2.50	4350	0.221	0.0284
6.72	3.71	2360	0.702	0.0282
10.20	7.36	1410	2.10	0.0279
3.70	2.19	3960	0.241	0.0297
				Av: 0.0285 ± 0.0008

[a]At 27.3°C in ethanol; data from Ref. 5.
[b]See Fig. 3-3.
[c]Calculated from Eq. (3-40).
[d]Assuming $R_{ex} = k[C_6H_5CH_2I][KI]$.

of the reaction mixture were added to benzene and water. The result was a separation of two layers, with potassium iodide contained in the aqueous phase and benzyl iodide in the organic. The activity of ^{131}I was determined in each layer. Data from one run are given in Table 3-3, and the semilogarithmic plot based on the McKay expression, Eq. (3-39), is shown in Fig. 3-3.

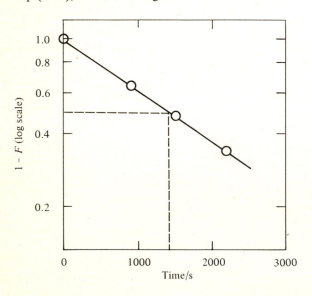

Figure 3-3 Semi-logarithmic plot of (1 – fraction exchange) against time for an exchange experiment between benzyl iodide and potassium iodide in ethanol at 27.3°C. See Table 3-3 and Ref. 5. The half-time for exchange is 1410 s.

A family of similar runs was also performed with different concentrations. Results of the experiments are also shown in Table 3-3, and they establish that the rate of exchange follows a mixed-second-order kinetic equation.

3-5 PARALLEL AND CONCURRENT FIRST-ORDER REACTIONS

A distinction can be made between a situation in which parallel reaction paths lead from the same reactants to the same products and that in which competing parallel reactions occur simultaneously and result in different products. Examples of both will be considered in subsequent sections.

In the first instance consider a single reagent A which reacts in independent reactions to produce different products:

$$\text{(3-41)}$$

The rate of consumption of A is

$$\frac{-d[A]}{dt} = (k_1 + k_2 + \cdots)[A] \tag{3-42}$$

which gives

$$[A] = [A]_0 \exp(-\Sigma k_n t) \tag{3-43}$$

where $\Sigma k_n = k_1 + k_2 + \cdots + k_n$. The rate of production of a given product P_i is

$$\frac{d[P_i]}{dt} = k_i[A] \tag{3-44}$$

Integrating after substitution for $[A]$ from Eq. (3-43) gives

$$[P_i]_t = \frac{k_i[A]_0}{\Sigma k_n} [1 - \exp(-\Sigma k_n t)] \tag{3-45}$$

The final yield of P_i is given by

$$[P_i]_\infty = \frac{k_i[A]_0}{\Sigma k_n} \tag{3-46}$$

Clearly, the various products P_i are produced in relative yields corresponding to the specific reaction rates, $k_1 : k_2 : k_3 : \ldots$. Product yield experiments serve to evaluate the ratio $k_i/\Sigma k_n$, as in Eq. (3-46). It is, of course, assumed that the products, once formed, are not interconvertible. The kinetic determination in which the concentration of the reactant is determined as a function of time provides a summation of the rate constants for all the independent reactions. That is, according to Eq. (3-43), the observed rate constant is Σk_n.

What is less obvious, and indeed startling when first encountered, is that the rate constant determined when any P_i is followed is also Σk_n and not k_i. This point can be developed as follows. Imagine that a spectrophotometric method is being used to monitor the reaction and that determinations are made at a wavelength at which only P_2 absorbs, the reactant and other products having $\epsilon = 0$. The absorbance-time relation, from Eq. (3-45), is

$$\text{Absorbance} = \frac{\epsilon_2 l k_2 [A]_0}{\Sigma k_n} [1 - \exp(-\Sigma k_n t)] \qquad (3\text{-}47)$$

$$\ln(\text{abs}_\infty - \text{abs}_t) = \ln \frac{\epsilon_2 l k_2 [A]_0}{\Sigma k_n} - \Sigma k_n t \qquad (3\text{-}48)$$

from which it is readily seen that the semilogarithmic plot suggested by Eq. (3-48) gives $k_{\text{obs}} = \Sigma k_n$ and not k_2. The rationale behind the result is that the analytical method is a monitor of the entire conversion—the rate at which P_2 is formed reflects the loss of A by all of the reactions which A undergoes. The time at which a given fraction of A has reacted is the same time as that at which the same fraction of the ultimate concentration of P_2 has formed.

A similar situation applies if reagent A reacts with B to form different products, as in Eq. (3-49). In this case an exactly analogous treatment applies, and the products

$$A + B \underset{k_2}{\overset{k_1}{\longrightarrow}} \begin{array}{c} P_1 \\ P_2 \\ \text{etc.} \end{array} \qquad (3\text{-}49)$$

are again formed in the ratio of the rate constants.

Closely related to these cases are ones in which the same products are formed from a single reactant but by parallel pathways. If such reactions have the same kinetic form, then, generally speaking, the existence of parallel pathways and an evaluation of their individual rate constants will not be revealed by these methods.

When the two pathways have different concentration dependences, however, the kinetic data will provide the desired resolution. As an example, consider the base hydrolysis of isopropyl bromide.[6] The reaction rate consists of two terms with respective first- and second-order components:

$$\frac{-d[(CH_3)_2CHBr]}{dt} = (k_1 + k_2[OH^-])[(CH_3)_2CHBr] \qquad (3\text{-}50)$$

The existence of (at least) two pathways is recognized from the form of the kinetic equation. The relative importance of each, which depends upon $[OH^-]$, is given by $k_1/k_2[OH^-]$. In fact, *three* pathways operate in parallel, because the bimolecular reaction itself consists of parallel substitution and elimination processes. The yields of isopropanol and propylene are used to ascertain the proportion of each pathway contributing to k_2, with due allowance for alcohol produced by the first-order pathway. (See Prob. 3-16.)

3-6 CONCURRENT REACTIONS OF MIXTURES

Suppose one added B to a mixture of substances A_1 and A_2, both of which react with B.

$$A_1 + B \xrightarrow{k_1} P_1 \tag{3-51}$$

$$A_2 + B \xrightarrow{k_2} P_2 \tag{3-52}$$

The rates of formation of the products are

$$\frac{d[P_1]}{dt} = k_1[A_1][B] \tag{3-53}$$

$$\frac{d[P_2]}{dt} = k_2[A_2][B] \tag{3-54}$$

Division gives

$$\frac{d[P_1]}{d[P_2]} = \frac{k_1}{k_2}\frac{[A_1]}{[A_2]} \tag{3-55}$$

Consider first the case when $[A_1]$ and $[A_2] \gg [B]$ and are then effectively constant, in which case the product ratio gives the rate constant ratio:

$$\frac{k_1}{k_2} = \frac{[P_1]_\infty[A_2]_0}{[P_2]_\infty[A_1]_0} \tag{3-56a}$$

In the case in which concentrations are not set such that $[A_1]$ and $[A_2]$ remain essentially constant during the course of the experiment, the relation becomes

$$\frac{k_1}{k_2} = \frac{\log\{[A_1]_0/([A_1] - [P_1]_\infty)\}}{\log\{[A_2]_0/([A_2] - [P_2]_\infty)\}} \tag{3-56b}$$

Use of Eq. (3-56) a and b permits determination of the relative rate ratios. Careful work requires experiments with several starting ratios of $[A_1]_0/[A_2]_0$.

These methods are particularly important in studying the reactivity of transient intermediates for which individual kinetic determinations are not feasible. The method has been widely applied, and one example should suffice. Bridger and Russell[7] have determined the relative rates of reaction of phenyl radicals with various hydrocarbons compared to CCl_4 [Eq. (3-57)]. (Phenyl radicals were generated by thermolysis of

$$C_6H_5 \cdot \diagup \begin{array}{l} \xrightarrow[k_H]{RH} C_6H_6 + R \cdot \\ \xrightarrow[k_{Cl}]{CCl_4} C_6H_5Cl + CCl_3 \cdot \end{array} \tag{3-57}$$

phenylazotriphenylmethane at $60°C$.) Their results for two hydrocarbons are shown in Fig. 3-4 via a plot of the product ratio $[C_6H_5Cl]/[C_6H_6]$ versus the reactant ratio $[CCl_4]/[RH]$. Equation (3-56a) was used because RH and CCl_4 were present in excess. The plots are linear through the origin, as expected, as shown in Fig. 3-4 for RH = toluene and cyclohexane.

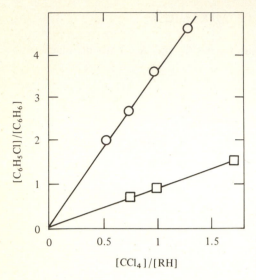

Figure 3-4 Competition experiments for reaction of the phenyl radical with RH and CCl_4, showing product ratio versus [Reactant] for the hydrocarbons toluene (○) and cyclohexane (□). Data from Ref. 7.

3-7 COMPETITION EXPERIMENTS FOR REACTION INTERMEDIATES

Two elegant methods for determining the reactivity of transient intermediates have been developed by Bunnett.[8] We shall confine ourselves to two illustrative examples of the methods.

The first instance, termed by Bunnett *ad eundem* competition (from the Latin, *ad eundum fructus*, to the same product), is illustrated by the reactions of benzyne with water and hydroxide ion. Both reactions form phenol:

$$C_6H_4 \diagup^{\xrightarrow[k_{H_2O}]{H_2O} C_6H_5OH}_{\xrightarrow[k_{OH^-}]{OH^-} C_6H_5OH} \tag{3-58}$$

If thiophenoxide ion is added as a third competing reagent, diphenylsulfide is formed:

$$C_6H_4 \xrightarrow[k_{PhS^-}]{C_6H_5S^-} (C_6H_5)_2S \tag{3-59}$$

The ratio of the yield of products is given by Eq. (3-60), which can be rearranged to

$$\frac{d[C_6H_5OH]}{d[(C_6H_5)_2S]} = \frac{k_{H_2O} + k_{OH^-}[OH^-]}{k_{PhS^-}[C_6H_5S^-]} = \frac{[C_6H_5OH]_\infty}{[(C_6H_5)_2S]_\infty} \tag{3-60}$$

the equation of a straight line [Eq. (3-61)]. These determinations thereby permit

$$\frac{[C_6H_5OH]_\infty[C_6H_5S^-]}{[(C_6H_5)_2S]_\infty} = \frac{k_{H_2O}}{k_{PhS^-}} + \frac{k_{OH^-}}{k_{PhS^-}}[OH^-] \tag{3-61}$$

evaluation of the desired reactivity ratio k_{OH^-}/k_{H_2O} as the ratio of slope to intercept. Without addition of a reagent such as the thiolate, that would not have been feasible. Other examples are given by Bunnett.[8]

The second method is termed *ad eosdem* competition, from the Latin *ad eosdem fructus*—to the same products (plural). An example similar to the preceding one is provided by the solvolysis of 4-chlorobenzyne by methanol and methoxide ions. Each parallel reaction itself consists of two parallel steps:

$$
\text{4-ClC}_6\text{H}_3
\begin{cases}
\xrightarrow[k_s^P]{\text{CH}_3\text{OH}} para\text{-ClC}_6\text{H}_4\text{OCH}_3 \\[4pt]
\xrightarrow[k_s^m]{\text{CH}_3\text{OH}} meta\text{-ClC}_6\text{H}_4\text{OCH}_3 \\[4pt]
\xrightarrow[k_b^P]{\text{CH}_3\text{O}^-} para\text{-ClC}_6\text{H}_4\text{OCH}_3 \\[4pt]
\xrightarrow[k_b^m]{\text{CH}_3\text{O}^-} meta\text{-ClC}_6\text{H}_4\text{OCH}_3
\end{cases}
\tag{3-62}
$$

The ratio of products is given (at constant $[\text{CH}_3\text{OH}]$ and $[\text{CH}_3\text{O}^-]$) by

$$
\frac{d[para]/dt}{d[meta]/dt} = \frac{[para]_\infty}{[meta]_\infty} = \frac{k_s^P[\text{CH}_3\text{OH}] + k_b^P[\text{CH}_3\text{O}^-]}{k_s^m[\text{CH}_3\text{OH}] + k_b^m[\text{CH}_3\text{O}^-]}
\tag{3-63}
$$

We further use the result that experiments done in acidified solution eliminate the methoxide ion terms and afford directly the ratio k_s^P/k_s^m. Equation (3-63) rearranges to the form

$$
\left(\frac{[para]_\infty}{[meta]_\infty} - \frac{k_s^P}{k_s^m}\right)\frac{[\text{CH}_3\text{OH}]}{[\text{CH}_3\text{O}^-]} = \frac{k_b^P}{k_s^m} - \frac{k_b^m}{k_s^m}\frac{[para]_\infty}{[meta]_\infty}
\tag{3-64}
$$

Hence, a plot of the left-hand side versus $[para]_\infty/[meta]_\infty$ should afford a straight line from whose slope and intercept the desired rate constant ratios can be determined.

Competition methods can provide only ratios of rate constants. For many purposes of comparison they will suffice. If it becomes feasible to measure any of the rate constants, however, the resolution of the competition results into actual rate constants becomes possible.

REFERENCES

1. Ewall, R. X., and D. H. Huchital: *Inorg. Chem.*, **14**:496 (1975).
2. King, E. L.: *J. Chem. Educ.*, **56**:580 (1979).
3. Swinehart, J. H., and G. W. Castellan: *Inorg. Chem.*, **3**:278 (1964).
4. McKay, H. A. C.: *Nature*, **142**:997 (1938); *J. Am. Chem. Soc.*, **65**:702 (1943).
5. Stillson, P., and M. Kahn: *J. Am. Chem. Soc.*, **75**:3580 (1953).
6. Hughes, E. D., and U. G. Shapiro: *J. Chem. Soc.*, 1177 (1937).
7. Bridger, R. F., and G. A. Russell: *J. Am. Chem. Soc.*, **85**:3754 (1963).
8. Bunnett, J. F.: in E. S. Lewis (ed.), "Investigations of Rates and Mechanisms of Reactions: General Considerations and Reactions at Conventional Rates," vol. VI, part 1 of the series "Techniques of Chemistry," Wiley-Interscience, New York, 1974, pp. 161–165.

PROBLEMS

3-1 *Opposing reactions.* D. S. Martin has studied the following reversible reaction spectrophotometrically.

$$Pt(NH_3)_3Cl^+ + Br^- \rightleftharpoons Pt(NH_3)_3Br^+ + Cl^-$$

At $\lambda = 285$ nm, a solution 1.00×10^{-3} M $Pt(NH_3)_3Cl^+$ in 0.2 M KCl in a 5-cm cell has an absorbance of 0.450. The molar absorptivity (extinction coefficient) of $Pt(NH_3)_3Br^+$ is 240 M^{-1} cm^{-1}.

The reaction was studied at 25°C (data given below), under these concentrations: $[Pt(NH_3)_3Cl^+]_0 = 1.00 \times 10^{-3}$, $[Cl^-] = 0.200$, $[Br^-] = 0.0500$ M. Assuming the reaction is an elementary one, evaluate the forward and reverse rate constants at 25.0°. (Save the results for Prob. 6-1.)

Time/s	0	600	1200	1800	3600	18000(∞)
Absorbance, 285 nm, b = 5 cm	0.450	0.550	0.632	0.688	0.784	0.843

3-2 *Reversible reactions.* Consider an experiment on the reversible first-order reaction (3-1) in which two identical solutions were prepared at different times. Designate the difference as τ. The first (older) solution was placed in the reference beam of a double-beam spectrophotometer and the second solution in the sample beam. The spectrophotometer records the net absorbance $D'(D' = D_2 - D_1)$ as a function of time.

Derive an equation showing how D' is related to the reaction kinetics. What is the slope of $\log D'$ versus time? (If necessary, consult Ref. 5b of Chap. 2.)

3-3 *Reversible reactions.* The kinetics of the reversible equilibrium

$$A + B \underset{-1}{\overset{1}{\rightleftharpoons}} C$$

was studied with [B] $\gg$ [A] or [C] by using an instrumental method to record a property P which is linear in concentration. A plot of $\ln[(P_t - P_{eq})/(P_0 - P_{eq})]$ versus t was linear; its slope is designated $-k_{obs}$.

From the following results compute k_1, k_{-1}, and the missing value of k_{obs}:

$[A]_0$/M	1.00×10^{-3}	4.00×10^{-3}	1.60×10^{-3}	3.20×10^{-3}
$[B]_0$/M	6.00×10^{-2}	1.2×10^{-1}	3.00×10^{-2}	9.00×10^{-2}
$[C]_0$/M	5.0×10^{-4}	0	1.0×10^{-3}	0
k_{obs}/s^{-1}	?	3.26×10^{-2}	1.70×10^{-2}	2.72×10^{-2}

3-4 *Opposing reactions.* A paper dealing with the kinetics of the NO_2-catalyzed isomerization of olefins [*J. Am. Chem. Soc.*, **96**:6549 (1974)] reports the following kinetic equation. Prove or disprove the expression given for k_{cis} in experiments starting only with *cis*- olefin.

$$trans\text{-Olefin} + NO_2 \underset{k_{cis}}{\overset{k_{trans}}{\rightleftharpoons}} NO_2 + cis\text{-olefin}$$

$$k_{cis}t = \frac{-1}{(1 + K_{TC})[NO_2]_0} \times \ln\left[1 - (1 + K_{TC})\frac{[T]_t}{[C]_0 + [T]_0}\right]$$

with $K_{TC} = k_{trans}/k_{cis}$.

3-5 *Parallel and reversible reactions.* The isomerization of allyl phenyl sulfide [*J. Am. Chem. Soc.*, **99**:3441 (1977)] is a degenerate rearrangement made detectable by isotopic labeling of an end of the allyl grouping, permitting kinetic monitoring by nmr techniques.

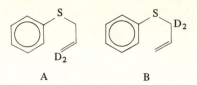

A   B

The isomerization of A to B was directly dependent on the initial concentration of A even though the apparent first-order plot was linear. The authors propose competing unimolecular and bimolecular processes, and they show that the system reduces to a first-order expression when the equilibrium constant K is unity.

$$A \underset{-1}{\overset{1}{\rightleftarrows}} B \qquad 2A \underset{-2}{\overset{2}{\rightleftarrows}} 2B$$

$$\frac{-d[A]}{dt} = k_1[A] - k_{-1}([A]_0 - [A]) + 2k_2[A]^2 - 2k_2([A]_0 - [A])^2$$

(a) Prove or disprove the final kinetic relation given:

$$-\ln([A] - [A]_{eq}) = 2(k_1 + k_2[A]_0)t$$

(b) What does occurrence of the following equilibration demonstrate?

$$PhSCH_2\underset{CH_3}{\overset{|}{C}}=CH_2 + p\text{-}CH_3\text{ØSCH}_2CH=CH_2 \rightleftharpoons PhSCH_2CH=CH_2 + p\text{-}CH_3\text{ØSCH}_2\underset{CH_3}{\overset{|}{C}}=CH_2$$

3-6 Exchange reactions. Electron exchange between Fe^{2+} and Fe^{3+} was studied by using ^{59}Fe radiotracer counting the activity of $Fe(OH)_3(s)$ precipitated after complexing Fe^{2+} as $Fe(phen)_3^{2+}$. Use the data given below to determine (a) $t_{1/2}$, (b) R_{ex}, and (c) k_1 assuming a second-order rate law and (d) $t_{1/2}$ for the same experiment but with $[Fe^{2+}] = 5.0 \times 10^{-3}$ M. This run had 2.5×10^{-3} M Fe^{2+}, 5.0×10^{-3} M Fe^{3+}, 0.50 M $HClO_4$, $0.0°C$.

Time/s	0	30	60	90	120	150	180	210	∞
Activity of $Fe(OH)_3(s)$ counts s^{-1}	0	117	206	270	322	370	404	432	534

3-7 Exchange reactions. K. V. Krishnamurty and A. C. Wahl have found that $t_{1/2} = 306$ s for isotopic exchange of 0.086 M V^{2+} and 0.086 M V^{3+} (25°C, 1.0 M H^+, 2.0 M ionic strength). Compute: (a) R_{ex}, (b) the second-order rate constant assuming a bimolecular mechanism, and (c) $t_{1/2}$ in an experiment having 0.043 M V^{2+} and 0.129 M V^{3+}.

3-8 Parallel reactions. S. B. Hanna et al. [*J. Am. Chem. Soc.*, **96**:7222 (1974)] studied the kinetics of coupling a diazonium salt (D^+) with naphthol (N) to form an azo dye (Az), a reaction complicated by the spontaneous decomposition of D^+:

$$D^+ + N \xrightarrow{k_2} Az + H^+ \qquad D^+ \xrightarrow{k_1} \text{inert products}$$

Use the data below to calculate the rate constant k_2 in an experiment under these conditions: 0°C, $\mu = 0.45$ (KCl), $[N] = 1 \times 10^{-2}$ M, $[D^+]_0 = 1.0 \times 10^{-4}$ M.

10^{-3} time/s	8.33	10.89	14.15	16.99	19.08	24.36	∞
10^5 $[Az]_t$/M	2.00	2.45	2.90	3.25	3.60	4.23	6.40

3-9 *Parallel reactions.* The author of a paper dealing with the decomposition of $Co(NH_3)_5NO_2^{2+}$ [*J. Inorg. Nucl. Chem.*, **29**:2795 (1967)] found that both $Co(NH_3)_5H_2O^{3+}$ and Co^{2+} were produced in parallel first-order reactions:

$$Co(NH_3)_5NO_2^{2+}(A) + H_2O \overset{1}{\underset{2}{\lessgtr}} \begin{array}{l} Co(NH_3)_5H_2O^{3+}(B) + NO_2^- \\ Co^{2+}(aq)(C) + 5NH_3 + NO_2 \end{array}$$

He developed an *approximate* treatment of the kinetic data based on absorbance D data in a 1-cm optical path:

$$D_0 = [A]_0\epsilon_A$$

$$D_t \simeq [A]\epsilon_A + [B]\epsilon_B \quad (\text{since } \epsilon_C \simeq 0 \text{ at this } \lambda)$$

$$D'_\infty = [A]_0\epsilon_B$$

$$D_t - D'_\infty = [A]\epsilon_A - ([A]_0 - [B])\epsilon_B$$

(The quantity D'_∞ is the theoretical final absorbance if A were transformed entirely to B.) During the initial stages of reaction, $[C] \ll [A]$, and hence

$$D_t - D'_\infty \simeq [A](\epsilon_A - \epsilon_B)$$

Thus a plot of $\ln (D_t - D'_\infty)$ versus t will approximate linearity in the early stages and have as slope $-(k_1 + k_2)$.

You are to prove that an *exact* treatment of the kinetic data is feasible. Derive equations for $\ln (D_t - D_\infty)$ versus time, D_∞ being the true final absorbance. Do not include the requirement that $\epsilon_C = 0$, and show that your relation will hold through the entire course of the reaction.

3-10 *Competition experiments.* Boyle and Bunnett [*J. Am. Chem. Soc.*, **96**:1418 (1974)] have employed competition experiments to evaluate the relative reactivity of the para-nitrophenyl radical ($N\cdot$) toward CH_3OH and CH_3O^-. The competing reaction is that with iodobenzene according to the scheme:

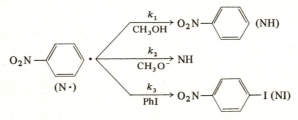

Assuming each reaction follows second-order kinetics, use the following data to evaluate k_1/k_2. (Four of their 20 experiments are given.)

$[CH_3O^-]$	0.254	0.418	0.762	1.05
$[PhI]$	0.250	0.500	0.250	0.500
$[NH]_\infty/[NI]_\infty$	2.09	1.23	3.27	2.06

1. $[N\cdot]_0 = 0.010$ M, so the $[CH_3O^-]$ and $[PhI]$ are effectively constant in any run.

2. CH_3OH is the solvent; in each run $a_{CH_3OH} = 1$.

3-11 *Parallel reactions.* P. Benson and A. Haim [*J. Am. Chem. Soc.*, **87**:3826 (1954)] have studied the rate of (1), complicated by (2), following $[SCN^-]$ produced at various times.

$$cis\text{-}Co(en)_2(NCS)Cl^+(=A) \overset{Fe^{2+}}{\underset{H_2O}{\lessgtr}} \begin{array}{l} Co^{2+} + Fe^{3+} + SCN^- + 2en + Cl^-(1) \\ cis\text{-}Co(en)_2(H_2O)NCS^+ + Cl^-(2) \end{array}$$

(a) Show that $\ln ([SCN^-]_\infty - [SCN^-]_t)$ versus time will be linear in experiments with $[Fe^{2+}] \gg [A]$. What rate constant(s) are given by the slope?

(b) Compute rate constant(s) from these data: an experiment with $[A]_0 = 9.1 \times 10^{-5}$ M and $[Fe^{2+}]_0 = 0.384$ M at $25.0°$ has $[SCN^-]_\infty = 7.8 \times 10^{-5}$ M and an apparent rate constant of 7.62×10^{-5} s^{-1}.

3-12 *Parallel reactions.* Consider a two-component mixture of substances A and B which react independently to form the same product C:

$$A \xrightarrow{k_A} C \quad \text{and} \quad B \xrightarrow{k_B} C$$

(a) Derive an equation for [C] as a function of time in a mixture with initial concentrations $[A]_0$ and $[B]_0$.

(b) Apply the relation of (a) to determine individual rate constants for the solvolysis of a mixture of the two chlorides $(CH_3)_3CC(Et)_2Cl$ and $(Et)_2CH_3CC(Me)_2Cl$, which are produced together in a synthesis. The kinetic data of H. C. Brown and R. S. Fletcher [*J. Am. Chem. Soc.*, 71:1845 (1949)] gave these values for the concentration of HCl as a function of time:

Time/s	0	1800	3600	5400	9000	10,800	12,600
C_{HCl}/M	0	0.01979	0.0293	0.0350	0.0431	0.0470	0.0492

Time/s	16,200	19,800	21,600	∞
C_{HCl}/M	0.0513	0.0534	0.0542	0.0580

(c) What is the initial concentration of each? How can one know which compound is which?

3-13 *Concentration-jump method.* The kinetics of the following disproportionation equilibrium has been studied by using the concentration-jump technique [T. Matsuda, K. Tanaka, and T. Tanaka, *Inorg. Chem.*, 18:454 (1979)]:

$$Mo_2O_3L_4 \underset{-1}{\overset{1}{\rightleftharpoons}} MoOL_2 + MoO_2L_2$$

These authors state: "When a 1,2-dichloroethane solution of [the complexes] was mixed with the same volume of 1,2-dichloroethane, the intensity of the [visible absorption] bands of the complexes decreased exponentially with half-lives of 0.1–1.8 s." The original solutions were prepared from only $Mo_2O_3L_4$ and no other Mo compound.

(a) Derive an expression which relates k_{obs} of the experiment to the equilibrium concentration(s) of the species involved. Assume that the shift in equilibrium is small.

(b) Show that the equation can be recast in a form in which k_{obs}^2 varies linearly with $[Mo_2O_3L_4]_0$, the latter being the *initial* concentration of the dinuclear complex after dilution, as if all species were present as this form. Use the result that, for $L = Et_2NCSe_2$, the equation of the straight line is:

$$k_{obs}^2 = 116 \text{ s}^{-2} + 4.0 \times 10^5 \text{ mol dm}^{-3} \text{ s}^{-2} [Mo_2O_3L_4]_0$$

to evaluate k_1 and k_{-1}.

3-14 *Reversible reactions.* Consider the elementary reaction

$$A + B \rightleftharpoons C + D$$

in the case where the equilibrium constant is unity. Brown and Sutin [*J. Am. Chem. Soc.*, 101:883 (1979)] have found such a situation with reactions of ruthenium-ammine complexes. They give an integrated equation applicable when C and D are absent initially:

$$\ln (x_e - x) - \ln x_e = k\alpha t$$

where x and x_e are [C] and $[C]_\infty$ and α is a constant. Derive this equation and find the expression for α. The situation bears some relation to the corresponding second-order exchange reaction. Show how the integrated rate laws for each are related.

3-15 *Reversible reactions.* Prove that the integrated rate law for the reaction $A \underset{k_r}{\overset{k_f}{\rightleftarrows}} C + D$, when $[C]_0 = [D]_0 = 0$, is

$$\ln \frac{[A]_0^2 - [A]_\infty[A]}{[A]_0([A] - [A]_\infty)} = \frac{[A]_0 + [A]_\infty}{[A]_0 - [A]_\infty} k_f t$$

3-16 *Parallel pathways.* Isopropyl bromide solvolyzes in basic solution by three parallel pathways:

$$i\text{-PrBr} + H_2O \xrightarrow{k_{1s}} i\text{-PrOH} + H^+ + Br^-$$

$$i\text{-PrBr} + OH^- \xrightarrow{k_{2s}} i\text{-PrOH} + Br^-$$

$$i\text{-PrBr} + OH^- \xrightarrow{k_{2e}} CH_2{=}CHCH_3 + H_2O + Br^-$$

(a) Present expressions for $-d[i\text{-PrBr}]/dt$, $d[i\text{-PrOH}]/dt$, $d[C_3H_6]/dt$, and $[C_3H_6]_\infty/[i\text{-PrOH}]_\infty$ applicable to a solution with $[OH^-]_0 \gg [i\text{-PrBr}]_0$.

(b) Show how the rate constants can be evaluated from the following data: (1) the slope and intercept of the linear plot of $k_{obs}(= -d \ln [i\text{-PrBr}]/dt)$ versus $[OH^-]$ and (2) the slope and intercept of the plot of $[i\text{-PrOH}]_\infty/[C_3H_6]_\infty$ versus $[OH^-]^{-1}$.

FOUR

CONSECUTIVE REACTIONS AND REACTION INTERMEDIATES

A very important general class of reactions is that in which a product of one elementary step is the reactant in a subsequent step. Such a substance, produced in one reaction and consumed in another, is referred to as an *intermediate*. In some situations an intermediate may attain a concentration comparable with the concentrations of other reagents. Perhaps more common is the situation in which the intermediate is so reactive that it remains at trace concentration throughout. It may or may not be capable of direct detection, but its presence can be manifest through the kinetic equations. The steady-state approximation proves a useful way of treating such systems mathematically, and its applicability and limitations are considered in this chapter.

It is helpful to keep in mind that the useful rate expressions are those for the reaction rate as a function of the concentrations of bulk species. That is because bulk species are the substances for which a direct measure of concentration may be made and whose concentration may be controlled by direct variation. The rate of reaction, were it to be expressed as a function of the concentration of an intermediate, might be perfectly correct, but because that concentration will generally be immeasurable by direct methods, such an expression has but limited utility.

4-1 CONSECUTIVE FIRST-ORDER REACTIONS

Consider a reaction sequence consisting of two first-order (or pseudo-first-order) steps occurring in series:

$$A \xrightarrow{k_1} B \xrightarrow{k_2} C \qquad (4\text{-}1)$$

Such a sequence is an important one. Not only is it characteristic of radioactive decay but it is found in many chemical situations as well. Note, in particular, that the reverse reaction of B to A has been omitted; kinetic schemes in which it cannot be neglected are considered in Sec. 4-3.

The differential equations and their solutions are as follows:

$$\frac{-d[A]}{dt} = k_1[A] \tag{4-2}$$

$$[A] = [A]_0 \exp(-k_1 t) \tag{4-3}$$

$$\frac{d[B]}{dt} = k_1[A] - k_2[B] \tag{4-4}$$

Equation (4-3) is substituted into Eq. (4-4). Then each term is multiplied by $\exp(k_2 t)$, and the result is a relation in which exact differentials are obtained:

$$\exp(k_2 t)\frac{d[B]}{dt} + \exp(k_2 t)[B]k_2 = k_1[A]_0 \exp(k_2 - k_1)t \tag{4-5}$$

Integration of this relation with $[B] = 0$ at $t = 0$ and rearrangement yields:

$$[B] = \frac{k_1[A]_0}{k_2 - k_1}[\exp(-k_1 t) - \exp(-k_2 t)] \tag{4-6}$$

Since $[B]$ is represented as the difference of two exponential functions, it will, as expected, go through a maximum. By differentiation of Eq. (4-6) and the setting of $d[B]/dt = 0$, one can easily show that

$$[B]_{max} = [A]_0\left(\frac{k_2}{k_1}\right)^{k_2/(k_1 - k_2)} \tag{4-7}$$

$$t_{max} = \frac{\ln(k_2/k_1)}{k_2 - k_1} \tag{4-8}$$

C does not form immediately as A reacts, owing to the intervention of B. The delay in formation of C, termed an *induction period*, is related to the time required for B to reach its maximum concentration. The point of inflection in $[C]$ occurs exactly at the maximum for $[B]$, as seen by noting that $d^2[C]/dt^2 = k_2(d[B]/dt)$.

An interesting case occurs when in the reaction sequence $A \longrightarrow B \longrightarrow C$ both rate constants have the same value. The solutions derived above are inapplicable, because the expressions become indeterminate. The reader can show (Prob. 4-2) that the appropriate solution for that case is

$$[B] = k[A]_0 t \exp(-kt) \tag{4-9}$$

Longer sequences of first-order reactions, such as in a radiochemical decay series, have been treated by similar methods.[1-3] If the first step is much slower than the others, then the situation that develops is one in which each intermediate is present in an amount, relative to $[A]$, in inverse proportion to its rate constant. Thus for the simple two-step sequence of Eq. (4-1), when $k_1 \ll k_2$, Eq. (4-6) reduces to $[B] \cong [A]k_1/k_2$. In radiochemistry this is termed a *secular equilibrium*.

We turn next to the question of how to analyze data for such a reaction sequence when the reaction occurs in just two steps as in Eq. (4-1). It is referred to as a *biphasic process*. Of particular interest is the situation in which an instrumental method is used to follow the progress of the reaction. (The situation in which actual concentrations of A and B are determined is less complex, and it should be evident from what follows.) Suppose we have measured a property to which each substance contributes linearly. It might be the absorbance D of the solution, whose value at a given time is given by

$$D_t = \epsilon_A [A]_t + \epsilon_B [B]_t + \epsilon_C [C]_t \tag{4-10}$$

Substituting relations (4-3) and (4-6) into (4-10) gives

$$D_t = \epsilon_A [A]_t + \epsilon_B [B]_t + \epsilon_C ([A]_0 - [A]_t - [B]_t) \tag{4-11}$$

$$D_t = \epsilon_C [A]_0 + (\epsilon_A - \epsilon_C)[A]_0 \exp(-k_1 t)$$

$$+ (\epsilon_B - \epsilon_C) \frac{k_1 [A]_0}{k_2 - k_1} [\exp(-k_1 t) - \exp(-k_2 t)] \tag{4-12}$$

Since $D_\infty = \epsilon_C [A]_0$, this becomes

$$D_t - D_\infty = \alpha \exp(-k_1 t) + \beta \exp(-k_2 t) \tag{4-13}$$

where α and β are the constants

$$\alpha = \frac{(\epsilon_B - \epsilon_A)k_1 + (\epsilon_A - \epsilon_C)k_2}{k_2 - k_1} [A]_0 \tag{4-14}$$

$$\beta = \frac{(\epsilon_C - \epsilon_B)k_1 [A]_0}{k_2 - k_1} \tag{4-15}$$

A plot of $\log(D_t - D_\infty)$ versus time will, according to Eq. (4-13), consist of two added straight-line segments. The segments can be resolved, provided the k's are not too close to one another, as follows. Consider the case $k_1 > k_2$. The long-time linear portion will have a slope $-k_2$. Extrapolation of that line to t_0 affords β. One next computes a new difference Δ,

$$\Delta = D_t - D_\infty - \beta \exp(-k_2 t) = \alpha \exp(-k_1 t) \tag{4-16}$$

and a plot of $\log \Delta$ versus t affords k_1. Both rate constants are thus known. The value of β also provides, via Eq. (4-15), the unknown molar absorptivity of the intermediate B, all other quantities here being known.

An illustration of the method comes from kinetic data for the reaction of nitrous acid and hydrogen peroxide.[4] Figure 4-1 shows the trace of absorbance (optical density) against time. The data can be treated by computer methods.[5] For our purposes we shall consider the graphical treatment shown in Fig. 4-2. The plot of $\log(D_t - D_\infty)$ versus time consists of two portions. The linear segment at long times corresponds to a first-order rate constant of 0.0854 s^{-1}. The rate constant applicable at short times can be evaluated by using Eq. (4-16). The quantity Δ of that expression is the difference between the data points at short times and the extrapolation of the long-time linear portion. Then $\log \Delta$ is plotted against time to get the other rate constant, 3.76 s^{-1} in this case. The graphical procedure just described is illustrated in Fig. 4-2. Particular

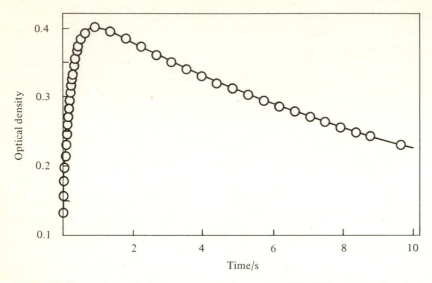

Figure 4-1 Formation and decay of peroxonitrous acid (HOONO) monitored at 320 nm [conditions pH 1.57, $T = 0.6°C$, $[HNO_2]_0 = 3.0 \times 10^{-3}$ M; $[H_2O_2] = 0.15$ M].[4] The solid line is that given by either set of equivalent parameters $k_1 = 3.76$ s^{-1}, $k_2 = 0.0854$ s^{-1}, $\epsilon_B = 5.83 \times 10^1$ dm^3 mol^{-1} cm^{-1}; $k_1' = 0.0854$ s^{-1}, $k_2' = 3.76$ s^{-1}, $\epsilon_B' = 2.039 \times 10^3$ dm^3 mol^{-1} cm^{-1}.

care is required in the extrapolation and in the computation of Δ to avoid accumulation of error.

It is tempting to interpret the findings in this case in terms of a fast reaction forming the peroxonitrous acid intermediate ($k = 3.76$ s^{-1}) and a slower reaction ($k = 0.0854$ s^{-1}) in which it is consumed. As we shall see in the next section, however, that is but one of two possible solutions.

Before leaving the present discussion, however, it is necessary to point out how easily a clean first-order reaction may be mistaken for a two-stage process. Consider a biphasic reaction followed spectrophotometrically at a wavelength at which the intermediate is not seen as prominently as it was in this case. For example, imagine that the nitrous acid–hydrogen peroxide reaction were followed at some wavelength at which the molar absorptivity of the intermediate was between that of the reactant and that of the product. In that case the absorbance (optical density) would rise or fall smoothly with time. The indication of the biphasic nature of the reaction would come only from the nonlinear pseudo-first-order kinetic plot. Particularly if the two rate constants were closer in value than these two are, the biphasic reaction might well appear as a first-order process in which the kinetic plot was curved because a small difference between the "correct" and observed values of the end-point reading P_∞ had been introduced.

Since the latter situation is all too common, more common than the occurrence of biphasic reactions, one must be very cautious in ascribing curved first-order kinetic plots to the occurrence of a biphasic pattern. A useful guide often comes from replicate experiments in which a different absorption band is monitored. If there are one or

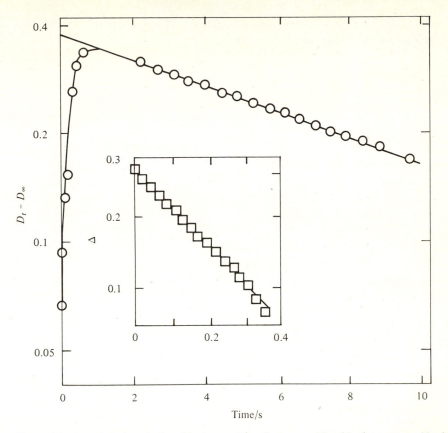

Figure 4-2 Plot of the kinetic data for the reaction of nitrous acid and hydrogen peroxide shown in Figure 4-1. The plot shows $D_t - D_\infty$ (log scale) versus time; the slope at long times gives $k = 0.0854$ s^{-1}. The inset shows the plot of log Δ [defined by Eq. (4-16)] versus time, and corresponds to $k = 3.76$ s^{-1}.

more isosbestic points in the spectrum, wavelengths at which two species have identical molar absorptivities, a biphasic kinetic pattern is most easily recognized and the rate constants are evaluated without recourse to the resolution of the sum of exponential terms. The cases are:

$\epsilon_B = \epsilon_C$ First-order throughout, with $k_{obs} = k_1$

$\epsilon_A = \epsilon_B$ First-order after an induction period, with $k_{obs} = k_2$

$\epsilon_A = \epsilon_C$ $t_{max} = \ln(k_1/k_2)/(k_1 - k_2)$

4-2 DUAL SOLUTIONS IN CONSECUTIVE REACTIONS

It is now recognized[4,6] that the kinetic situation $A \longrightarrow B \longrightarrow C$ admits of two mathematical solutions. There will always be a set of alternative parameters, k_1', k_2', and ϵ_B', which will fit the data with identical precision. The sets of rate constants are such that the fast and slow kinetic steps are interchanged:

$$k_1 = k_2' \quad \text{and} \quad k_2 = k_1' \tag{4-17a}$$

The alternative molar absorptivity is given by

$$\epsilon_B' = \epsilon_A + \frac{k_1(\epsilon_B - \epsilon_A)}{k_2} \tag{4-17b}$$

a relation that is readily derived by substitution of Eq. (4-17a) into (4-13) and (4-15). Alcock et al.[4] summarize the situation: "An optical density which rapidly increases and slowly declines does not necessarily imply a fast first and slow second reaction." They cite an example of such an ambiguity in the reaction of nitrous acid and hydrogen peroxide, which proceeds by way of a peroxonitrous acid intermediate (Fig. 4-1).

How does one determine which is the correct set of parameters? Alcock et al.[4] suggest that if k_1 and k_2 are capable of independent variation (as will be the case if either one is a pseudo-first-order constant incorporating a variable excess concentration), then the question is answered by whether ϵ_B or ϵ_B' remains independent of such variation. [Note that Eq. (4-17b) relates the two molar absorptivities by the ratio k_1/k_2.] Other authors[7] choose between the formulations based on the reasonableness of the two values of ϵ_B versus ϵ_B'. The values are often vastly different, permitting a choice between the assignments if the intermediate is a substance for which stable analogs of known absorption spectra are known.

Consider the circumstances under which an authentic biphasic reaction will resemble a single-stage reaction and give a linear plot of log $(D_t - D_\infty)$ versus time.[6,8] There are the trivial cases

$$k_1 \ll k_2 \quad (\text{thus } k_{obs} = k_1) \tag{4-18a}$$

$$k_1 \gg k_2 \quad (\text{thus } k_{obs} = k_2) \tag{4-18b}$$

It was shown earlier (Sec. 4-1) that linear first-order plots will also result, and may provide considerable assistance in evaluation of the rate constants, if the data are measured at an isosbestic point. An intriguing situation is the one in which the rate constants and molar absorptivities are related in such a way that the plot becomes linear. Such a condition will result on rare occasion by accident, but it may occur (and persist at many wavelength bands) when a polyfunctional compound reacts sequentially. That is likely to happen provided the successive rate constants and molar absorptivities are in exact statistical ratio.[9] Regardless of its origin, the condition required to cause the first-order plot to be linear for a biphasic reaction is

$$\frac{k_1}{k_2} = \frac{\epsilon_C - \epsilon_A}{\epsilon_B - \epsilon_A} \tag{4-19}$$

An example[8] is found in the reaction in Eq. (4-20) which was studied under conditions such that $\epsilon_A = 0$. The condition

$$[Mo(Cp)(CO)_3(SnMe_3)] \xrightarrow{I_2} complex \tag{4-20}$$

A — complex; B

$$complex \longrightarrow [Mo(Cp)(CO)_3I] + Me_3SnI$$

B — C

for an apparent single stage simplifies to

$$\frac{k_2}{k_1} = \frac{\epsilon_B}{\epsilon_C} \tag{4-21}$$

The value of k_{obs} is $k_1\epsilon_B/\epsilon_C$, which becomes k_2 in the limit where Eq. (4-21) applies. The kinetic relation is thus (in this special case)

$$D_t - D_\infty = -[A]_0\epsilon_C \exp(-k_2t) \tag{4-22}$$

For data of the customary precision, the kinetic plot may be satisfactorily linear even if Eq. (4-21) is only approximately obeyed. That may lead to a seeming wavelength dependence of the apparent rate constant, $k_{obs} \sim \epsilon_B k_1/\epsilon_C$.

4-3 A CONSECUTIVE REACTION WITH A REVERSIBLE STEP

The reaction sequence for the net transformation of A to C via reversible formation of intermediate B, Eq. (4-23), is of interest in its own right. Consider first the general

$$A \underset{k_{-1}}{\overset{k_1}{\rightleftharpoons}} B \xrightarrow{k_2} C \tag{4-23}$$

solution. The differential rate equations are

$$\frac{-d[A]}{dt} = k_1[A] - k_{-1}[B] \tag{4-24}$$

$$\frac{d[B]}{dt} = k_1[A] - k_{-1}[B] - k_2[B] \tag{4-25}$$

$$\frac{d[C]}{dt} = k_2[B] \tag{4-26}$$

The solutions (for $[B]_0 = 0$ and $[C]_0 = 0$) are

$$[A] = \frac{k_1[A]_0}{\lambda_2 - \lambda_3}\left[\frac{\lambda_2 - k_2}{\lambda_2}\exp(-\lambda_2 t) - \frac{\lambda_3 - k_2}{\lambda_3}\exp(-\lambda_3 t)\right] \tag{4-27}$$

$$[B] = \frac{k_1[A]_0}{\lambda_2 - \lambda_3}\left[\exp(-\lambda_3 t) - \exp(-\lambda_2 t)\right] \tag{4-28}$$

where $\lambda_2 = \frac{1}{2}(p + q)$ and $\lambda_3 = \frac{1}{2}(p - q)$ with $p = k_1 + k_{-1} + k_2$ and $q = (p^2 - 4k_1k_2)^{1/2}$. The expression for [C] is easily derived as $[A]_0 - [A] - [B]$.

Application of the general solution can be unwieldy,[10,11] and it is thus useful to examine certain special cases. The most notable of them is the steady-state approximation considered in the next section. Another is the case $k_{-1} \approx 0$, already treated in Sec. 4-1.

4-4 THE STEADY-STATE APPROXIMATION

The reaction sequence of Eq. (4-23) leads, in the general case, to a cumbersome solution [Eqs. (4-27) and (4-28)] which simplifies in certain circumstances to a far more tractable form. The simplification to be considered, known as the *steady-state* (or *stationary-state*) *approximation*, is useful when the intermediate B is so reactive that it does not accumulate to an appreciable level compared with A and C. The situation is a common one, and the method finds wide application in many different chemical systems.

As it is usually stated, the method is said to entail setting to zero the first derivative of the concentration of each of the intermediates. As we shall see, that is an unduly strict statement of the condition which must be met to validate the method. Nonetheless, proceeding in this fashion for the moment and equating $d[B]/dt$ to zero, Eq. (4-25) can be rearranged to give

$$[B]_{ss} = \frac{k_1[A]}{k_{-1} + k_2} \tag{4-29}$$

The concentration of the intermediate is shown here with an explicit label identifying the steady-state approximation.

Substitution of this expression into Eqs. (4-24) and (4-26) yields

$$\frac{-d[A]}{dt} = k_1[A] - \frac{k_{-1}k_1[A]}{k_{-1} + k_2} = \frac{k_1 k_2[A]}{k_{-1} + k_2} \tag{4-30}$$

$$\frac{d[C]}{dt} = k_2[B] = \frac{k_1 k_2[A]}{k_{-1} + k_2} \tag{4-31}$$

The result that $-d[A]/dt$ and $d[C]/dt$ are equal is a consequence of setting $d[B]/dt = 0$, since conservation of mass requires

$$[A] + [B] + [C] = [A]_0 \tag{4-32}$$

Actually, of course, $d[B]/dt$ is not zero. If it were, according to Eq. (4-29), $d[A]/dt$ would also be zero and no reaction would occur. Thus it is worthwhile to examine more closely the nature of this approximation which is so widely useful.

It is a necessary and sufficient condition for the validity of these relations that [B] be $\ll [A] + [C]$. We can see that is the case by comparing the material balance expressions for [A] and [C]:

$$[A] = [A]_0 - [C] - [B] \tag{4-33}$$

and
$$[C] = [A]_0 - [A] - [B] \tag{4-34}$$

If, indeed, $[B] \ll [A] + [C]$, then

$$[A] + [C] = [A]_0 \tag{4-35}$$

which, upon differentiation, yields

$$\frac{-d[A]}{dt} = \frac{d[C]}{dt} \tag{4-36}$$

But we already know, by differentiation of Eq. (4-32), that

$$\frac{d[A]}{dt} + \frac{d[B]}{dt} + \frac{d[C]}{dt} = 0 \tag{4-37}$$

Comparison of Eq. (4-36) with Eq. (4-37) confirms that the condition $[B] \ll [A] + [C]$ is a sufficient and necessary requirement.

Another approach to exploring the nature of the steady-state approximation is based upon a rearrangement of the general (rigorous) rate law in Eq. (4-25) and yields

$$[B] = \frac{k_1[A] - d[B]/dt}{k_{-1} + k_2} \tag{4-38}$$

The expression for $-d[A]/dt$ becomes

$$\frac{-d[A]}{dt} = k_1[A] - \frac{k_{-1}(k_1[A] - d[B]/dt)}{k_{-1} + k_2}$$

$$\frac{-d[A]}{dt} = \frac{k_1 k_2[A]}{k_{-1} + k_2} + \frac{k_{-1}d[B]/dt}{k_{-1} + k_2} \tag{4-39}$$

This general equation contains a second term not found in the steady-state solution, Eq. (4-30). Comparison of the two shows that the last term in Eq. (4-39) must be much smaller than the first if the steady-state approximation is to apply. That gives another view of the condition needed to reduce the general solution to the simplified one, namely:

$$\frac{d[B]}{dt} \ll \frac{k_1 k_2[A]}{k_{-1}}$$

which is a condition far less restrictive than the stark mathematical statement $d[B]/dt = 0$. Other authors[11,12] have examined the conditions attending the steady-state approximation in this and related mechanisms.

Even though not an exact requirement, the condition $d[B]/dt = 0$ does provide a useful route for the derivation of rate laws. Keep in mind that the goal is to express the rate of reaction in terms of the concentrations of the bulk species, reactants, products, or other substances added in the reaction mixture, eliminating the concentrations of the reaction intermediate(s). For example, consider the classical S_N1 mechanism for the substitution reaction of an organic halide RX by a nucleophile Y^-:

$$RX \underset{-1}{\overset{1}{\rightleftharpoons}} R^+ + X^- \tag{4-40}$$

$$R^+ + Y^- \overset{2}{\longrightarrow} RY \tag{4-41}$$

The steady-state approximation, if applied to the concentration of the carbonium ion intermediate, gives

$$\frac{d[R^+]}{dt} = k_1[RX] - k_{-1}[R^+][X^-] - k_2[R^+][Y^-] = 0 \qquad (4\text{-}42)$$

Solution for $[R^+]$ gives

$$[R^+]_{ss} = \frac{k_1[RX]}{k_{-1}[X^-] + k_2[Y^-]} \qquad (4\text{-}43)$$

and the reaction rate becomes

$$\frac{d[RY]}{dt} = \frac{k_1 k_2[RX][Y^-]}{k_{-1}[X^-] + k_2[Y^-]} \qquad (4\text{-}44)$$

It is often important to compare the form of an experimental rate law with the one derived for proposed mechanisms. The comparison is needed to learn whether each postulated mechanism is consistent with the data. For that reason one must develop some facility with the derivations. Particularly if the mechanism is a complex one, a complete mathematical analysis starting with $d[B]/dt = 0$ can be tedious. It is thus useful to consider some algebraic shortcuts which reduce the manipulations required. One of them, applicable when all the reactions of the intermediate show a first-order dependence upon its concentration, is to write the steady-state concentration directly by use of the following relation:

$$[\text{Int}] = \frac{\text{sum of the rates of all the steps producing the intermediate}}{\text{sum of all the rate constants, rates/[int], destroying the intermediate}} \qquad (4\text{-}45)$$

This recipe is a very easy one to apply. For the S_N1 mechanism of Eqs. (4-40) and (4-41) the concentration of the intermediate is immediately given by

$$[R^+]_{ss} = \frac{k_1[RX]}{k_{-1}[X^-] + k_2[Y^-]} \qquad (4\text{-}46)$$

This expression is identical with that given by Eq. (4-43), and the reaction rate, Eq. (4-44), follows directly.

A slightly more complex example concerns a hydrolysis step in competition with capture of R^+ by Y^-. That is to say, in aqueous solution, the mechanism may be

$$RX \underset{-1}{\overset{1}{\rightleftharpoons}} R^+ + X^- \qquad (4\text{-}40)$$

$$R^+ + Y^- \overset{2}{\longrightarrow} RY \qquad (4\text{-}41)$$

$$R^+ + H_2O \overset{3}{\longrightarrow} ROH + H^+ \qquad (4\text{-}47)$$

From Eq. (4-45), $[R^+]_{ss}$ is immediately given as

$$[R^+]_{ss} = \frac{k_1[RX]}{k_{-1}[X^-] + k_2[Y^-] + k_3} \qquad (4\text{-}48)$$

[Note that the rate of reaction (4-47) has been written $k_3[R^+]a_{H_2O}$ with $a_{H_2O} = 1$, in an aqueous solution.] The rate of reaction is thus

$$\frac{d[RX]}{dt} = \frac{k_1[RX](k_2[Y^-] + k_3)}{k_{-1}[X^-] + k_3 + k_2[Y^-]} \tag{4-49}$$

which can also be seen to be the sum of the rates of formation of the individual products:

$$\frac{d[ROH]}{dt} = \frac{k_1 k_3[RX]}{k_{-1}[X^-] + k_3 + k_2[Y^-]} \tag{4-50}$$

$$\frac{d[RY]}{dt} = \frac{k_1 k_2[RX][Y^-]}{k_{-1}[X^-] + k_3 + k_2[Y^-]} \tag{4-51}$$

As a final example of the algebra of steady-state intermediates, consider a mechanism in which two intermediates occur. For example, consider that an ion-pair intermediate, $R^+ \cdot X^-$ is formed prior to the separated ions. Considering hydrolysis only, the mechanism being examined is

$$RX \underset{-1}{\overset{1}{\rightleftharpoons}} R^+ \cdot X^- \tag{4-52}$$

$$R^+ \cdot X^- \underset{-2}{\overset{2}{\rightleftharpoons}} R^+ + X^- \tag{4-53}$$

$$R^+ + H_2O \overset{3}{\longrightarrow} ROH + H^+ \tag{4-54}$$

The intermediates have, according to Eq. (4-45), steady-state concentrations given by

$$[R^+ \cdot X^-]_{ss} = \frac{k_1[RX] + k_{-2}[R^+]_{ss}[X^-]}{k_{-1} + k_2} \tag{4-55}$$

$$[R^+]_{ss} = \frac{k_2[R^+ \cdot X^-]_{ss}}{k_{-2}[X^-] + k_3} \tag{4-56}$$

Further manipulation is required, however, since these expressions give one intermediate expressed in terms of the other. Combination of the two gives

$$[R^+]_{ss} = \frac{k_1 k_2[RX]}{k_2 k_3 + k_{-1} k_3 + k_{-2} k_3[X^-]} \tag{4-57}$$

and the reaction rate in the steady state is

$$\frac{d[ROH]}{dt} = \frac{k_1 k_2 k_3[RX]}{k_2 k_3 + k_{-1} k_3 + k_{-2} k_3[X^-]} \tag{4-58}$$

More needs to be said about these equations, particularly about the significance of the multiple-denominator terms. We shall return to this point following a consideration of the concept of rate-limiting step.

4-5 LIMITING FORMS; THE RATE-LIMITING STEP

The steady-state rate law for the reaction sequence

$$A \underset{-1}{\overset{1}{\rightleftharpoons}} B \overset{2}{\longrightarrow} C \tag{4-23}$$

which is given by Eqs. (4-30) and (4-31) clearly has two extreme forms, depending upon the magnitudes of k_{-1} and k_2. Thus in a system with $k_{-1} \ll k_2$, the rate is

$$\frac{-d[A]}{dt} \cong k_1[A] \tag{4-59}$$

As the form of Eq. (4-59) clearly indicates, this corresponds to the case in which the first step is the slower of the two. In such circumstances the conversion of A into B is termed the *rate-limiting step*. Note that the condition by which Eq. (4-59) was obtained concerns the relative reactivity of intermediate B. The condition $k_{-1} \ll k_2$ indicates that most molecules of B react to form C; only a very small proportion reverts to A.

The case with $k_{-1} \gg k_2$ leads to the limiting form

$$\frac{-d[A]}{dt} = \frac{k_1}{k_{-1}} k_2[A] \tag{4-60}$$

The limiting form corresponds to the production of a small equilibrium concentration of B in the first reaction and the rate-limiting conversion of B to C. The probable reaction of B is reversion to A; only an occasional molecule of B reacts to produce C. In this sequence the limits correspond to those set by the relative values of k_{-1} and k_2. These are constants of the system, and under a given set of experimental conditions (solvent, pH, temperature, etc.) they are beyond variation by the experimenter.

For a similar sequence, but one with additional chemical species, the two limiting forms are often more easily realized in practice. The reaction of Tl(III) and Fe(II) [Eq. (4-61)] is believed to proceed

$$2Fe(II) + Tl(III) = 2Fe(III) + Tl(I) \tag{4-61}$$

by two consecutive steps involving an intermediate Tl(II):

$$Fe(II) + Tl(III) \underset{-1}{\overset{1}{\rightleftharpoons}} Fe(III) + Tl(II) \tag{4-62}$$

$$Fe(II) + Tl(II) \overset{2}{\longrightarrow} Fe(III) + Tl(I) \tag{4-63}$$

The rate law can be obtained by using the customary steady-state procedure, although the situation of two consecutive steps† arises often enough that a shorter method is advantageous. The net rate of reaction is equal to the rate at which the intermediate is produced multiplied by the fraction of its (the intermediate's) reactions which lead to

---

†This method applied to reactions of more than two steps will, however, lead to extraneous denominator terms. See Prob. 4-5.

product. Thus

$$\frac{d[\text{Tl(I)}]}{dt} = k_1 [\text{Fe(II)}] [\text{Tl(III)}] \times \frac{k_2 [\text{Fe(II)}]}{k_2 [\text{Fe(II)}] + k_{-1} [\text{Fe(III)}]} \qquad (4\text{-}64)$$

The two limiting forms are

(a) $k_{-1} [\text{Fe(III)}] \ll k_2 [\text{Fe(II)}]$,

$$\frac{d[\text{Tl(I)}]}{dt} \cong k_1 [\text{Fe(II)}] [\text{Tl(III)}] \qquad (4\text{-}65)$$

(b) $k_{-1} [\text{Fe(III)}] \gg k_2 [\text{Fe(II)}]$,

$$\frac{d[\text{Tl(I)}]}{dt} \cong \frac{k_1}{k_{-1}} k_2 \frac{[\text{Fe(II)}]^2 [\text{Tl(III)}]}{[\text{Fe(III)}]} \qquad (4\text{-}66)$$

As before, these two limiting cases correspond to the first and second reaction, respectively, being the rate-limiting step under the given circumstances. In this case, however, the experimenter plays a role in deciding the balance struck between the two denominator terms. It is the relative value of $k_{-1} [\text{Fe(III)}]$ versus $k_2 [\text{Fe(II)}]$ which is significant, and these concentrations can be varied over whatever range the experimental conditions and other considerations allow.

Return now to mechanisms considered in the section immediately preceding this. The two-step $S_N 1$ mechanism:

$$\text{RX} \underset{-1}{\overset{1}{\rightleftharpoons}} \text{R}^+ + \text{X}^- \qquad (4\text{-}40)$$

$$\text{R}^+ + \text{Y}^- \overset{2}{\longrightarrow} \text{RY} \qquad (4\text{-}41)$$

has a steady-state rate law easily derived by this shorter method:

$$\frac{d[\text{RY}]}{dt} = k_1 [\text{RX}] \times \frac{k_2 [\text{Y}^-]}{k_{-1} [\text{X}^-] + k_2 [\text{Y}^-]} \qquad (4\text{-}67)$$

This derivation is simpler still than that considered before. Under conditions in which $k_1 [\text{X}^-] \ll k_2 [\text{Y}^-]$, the rate is $d[\text{RY}]/dt \simeq k_1 [\text{RX}]$ and the first step is rate-limiting. When the reverse inequality applies, the rate is $d[\text{RY}]/dt \simeq (k_1 k_2 / k_{-1})[\text{RX}][\text{Y}^-]/[\text{X}^-]$. In effect, the first step is at equilibrium, with $[\text{R}^+] \simeq k_1 [\text{RX}]/k_{-1}[\text{X}^-]$, and the second step is rate-limiting.

If the three-step sequence involving an ion pair and a carbonium ion intermediate is similarly analyzed, Eqs. (4-52) to (4-54), three limiting forms are obtained: $k_1 [\text{RX}]$, $(k_1 k_2 / k_{-1})[\text{RX}]$, and $(k_1 k_2 k_3 / k_{-1} k_{-2})[\text{RX}]/[\text{X}^-]$. Consideration of these forms shows that each in turn corresponds to the first, second, and third steps of the mechanistic sequence being rate-limiting.

A corollary of this is that the number of terms in the denominator of a complete rate law corresponds to the number of potential rate-limiting steps in the sequence. This generalization is useful in the interpretation of rate laws, and it will be considered further in Chap. 5.

4-6 DIRECT VERSUS SEQUENTIAL REACTION

How do we know whether the conversion of reactant(s) to product(s) proceeds via an intermediate? A question of that sort does not admit of a general answer, but it is useful to consider the same example as before, the net transformation of A to C, possibly by way of steady-state intermediate B.

$$A \underset{-1}{\overset{1}{\rightleftarrows}} B \overset{2}{\longrightarrow} C \qquad (4\text{-}23)$$

In the steady state the reaction is first-order with respect to [A]:

$$\frac{-d[A]}{dt} = \frac{d[C]}{dt} = \frac{k_1 k_2}{k_{-1} + k_2} [A] \qquad (4\text{-}30)$$

If, instead, A is transformed directly to C,

$$A \overset{k'}{\longrightarrow} C \qquad (4\text{-}68)$$

the reaction is also first-order:

$$\frac{-d[A]}{dt} = \frac{d[C]}{dt} = k'[A] \qquad (4\text{-}69)$$

The kinetics do not distinguish these two cases, because a first-order dependence upon [A] is found for both. This result is not an uncommon one in kinetics; one mechanism may be consistent with a rate expression, but other mechanisms may also lead to the same rate expression.

If it proved possible to detect substance B during the course of the reaction of A to form C, say by UV or esr spectroscopy, would that establish B was an intermediate? The answer is negative:

$$A \underset{-1}{\overset{1}{\rightleftarrows}} B \qquad (4\text{-}70)$$

$$A \overset{k'}{\longrightarrow} C \qquad (4\text{-}71)$$

$$\frac{-d[A]}{dt} = k'[A] \qquad (4\text{-}72)$$

First-order kinetics apply to this mechanism as well as mechanisms in which B is a steady-state intermediate. In other words, B may exist in the system without being an intermediate; the reaction may proceed in spite of its presence and not because of it.

An effective way to learn whether B is an intermediate is to find a substance X with which it reacts very rapidly, preventing B's returning to A or a further reaction to form C. If B is an intermediate, $d[BX]/dt$ will be $k_1[A]$, which may or may not be identical with k_{exp}. (If it is, then that establishes the intermediacy of B in the scheme $A \rightleftarrows B \rightarrow C$.) If a scheme $A \rightleftarrows B$, $A \rightarrow C$ is correct, $d[BX]/dt$ will be $k_1[A]$, whereas $-d[A]/dt$ is $k'[A]$.

Whether a further distinction can be made in a particular case depends on the numerical values. The first scheme has $k_{app} = k_1/(1 + k_{-1}/k_2)$. If B is then to be identified as an intermediate, $d[BX]/dt$ may be no faster than $d[B]/dt$, which has an upper limit of $k_1[A]$. In other words, if $k_{app} = [A]^{-1}d[BX]/dt$, the first mechanism is strongly suggested. If $k_{app} < [A]^{-1}d[BX]/dt$, either mechanism may hold. But if $k_{app} > [A]^{-1}d[BX]/dt$, the first alternative is eliminated.

4-7 AN INTERMEDIATE WHICH REACTS IN STEPS HAVING DIFFERENT REACTION ORDERS

An interesting situation arises when the competing reactions of an intermediate in a reaction sequence show different reaction orders with respect to the concentration of the intermediate, such as one reaction having a first-order dependence and the other a second-order dependence. Then the steady-state approximation leads to a quadratic equation. One example is the following general scheme in which A_2 represents a symmetrical species which undergoes unimolecular dissociation.

$$A_2 \underset{-1}{\overset{1}{\rightleftharpoons}} 2A \tag{4-73}$$

$$A \overset{2}{\longrightarrow} P \tag{4-74}$$

In this scheme A represents an intermediate which reacts bimolecularly to re-form reactant A_2 or unimolecularly to yield the product; included implicitly in this general mechanism are schemes with a second bimolecular step run under pseudo-first-order conditions, i.e., $A + B \overset{k_2'}{\longrightarrow} P'$ with $[B] \gg [A_2]$ and $[B]$ thus effectively constant.

The steady-state expression for $[A]$ gives the relation

$$2k_{-1}[A]^2 + k_2[A] - 2k_1[A_2] = 0 \tag{4-75}$$

The quadratic formula provides an expression for $[A]$; we need take one root only, since $[A]$ must be positive:

$$[A]_{ss} = \frac{-k_2 + \sqrt{k_2^2 + 16k_1 k_{-1}[A_2]}}{4k_{-1}} \tag{4-76}$$

The notation is simplified by inclusion of a composite constant a defined by

$$a = \frac{16k_1 k_{-1}}{k_2^2} \tag{4-77}$$

and the steady-state solution thus becomes

$$[A]_{ss} = \frac{\sqrt{1 + a[A_2]} - 1}{k_2 a/4k_1} \tag{4-78}$$

The rate of reaction is

$$\frac{-d[A_2]}{dt} = \frac{1}{2}\frac{d[P]}{dt} = k_1[A_2] - k_{-1}[A]^2 = \frac{k_2}{2}[A] \tag{4-79}$$

Substitution of [A] affords an expression of the reaction rate

$$\frac{-d[A_2]}{dt} = \frac{2k_1}{a}(\sqrt{1 + a[A_2]} - 1) \tag{4-80}$$

It is useful to rearrange this expression to a different form, one in which the limiting kinetic forms are more readily evident and in which the numerator contains only the forward rate of the first step. Such a form is realized by multiplication of numerator and denominator of the right-hand side by $(\sqrt{1 + a[A_2]} + 1)$:

$$\frac{-d[A_2]}{dt} = \frac{2k_1(\sqrt{1 + a[A_2]} - 1)}{a} \times \frac{\sqrt{1 + a[A_2]} + 1}{\sqrt{1 + a[A_2]} + 1}$$

$$\frac{-d[A_2]}{dt} = \frac{2k_1[A_2]}{\sqrt{1 + a[A_2]} + 1} \tag{4-81}$$

Equation (4-81) is a useful form, since the two limits are immediately realized. Consider first the conditions under which $1 \gg a[A_2]$, which corresponds to conditions under which the first step is rate-limiting. The reaction rate reduces to:

$$\frac{-d[A_2]}{dt} = \frac{2k_1[A_2]}{1 + 1} = k_1[A_2] \tag{4-82}$$

The other limit occurs when $1 \ll a[A_2]$, and it is

$$\frac{-d[A_2]}{dt} = \frac{2k_1[A_2]}{\sqrt{a[A_2]}} = \frac{1}{2}\left(\frac{k_1}{k_{-1}}\right)^{1/2} k_2[A_2]^{1/2} \tag{4-83}$$

Again, this form is quite readily seen to apply when the second step is rate-limiting, [A] being simply its value when the first step is at equilibrium.

This particular solution is that given by authors[13] who found that the decomposition of $Mn_2(CO)_{10}$ follows such a mechanism, with the mononuclear manganese pentacarbonyl, $Mn(CO)_5$, functioning as the steady-state intermediate:

$$Mn_2(CO)_{10} \rightleftharpoons 2Mn(CO)_5 \tag{4-84}$$

$$Mn(CO)_5 \longrightarrow \text{decomposition products} \tag{4-85}$$

4-8 KINETIC EQUATIONS FOR CATALYZED AND ENZYME-CATALYZED REACTIONS

An additional term appears in the rate law for reactions accelerated by a trace concentration of a catalyst. To illustrate that, consider the following reaction scheme. Imagine first that E is not a catalyst at all, but is consumed in a stoichiometric reaction with substrate S:

$$E + S \underset{-1}{\overset{1}{\rightleftharpoons}} E \cdot S \tag{4-86}$$

$$E \cdot S \overset{2}{\longrightarrow} \text{products} \tag{4-87}$$

With the steady-state approximation for [E·S],

$$\frac{-d[S]}{dt} = -\frac{d[E]}{dt} = \frac{d[P]}{dt} = \frac{k_1 k_2 [E][S]}{k_{-1} + k_2} \tag{4-88}$$

If the experiments have $[S]_0 \gg [E]_0$, the reaction will follow first-order kinetics:

$$\frac{d[P]}{dt} = \frac{k_1 k_2 [S]}{k_{-1} + k_2} [E]_0 \tag{4-89}$$

Now, however, consider the situation if substance E is taken to represent a catalyst. Then the chemical identity of the second mechanistic step becomes

$$E \cdot S \xrightarrow{\ 2\ } P + E \tag{4-90}$$

In this event the catalyst E, present at very low concentration, remains constant, and the major substance S is consumed entirely.[†] In this event Eq. (4-90) also applies, but now the value for $[E]_0$ is no longer known. Account must be taken of the fact that an appreciable quantity of E may exist during the reaction complexed as E·S. (Of course, only a trivial quantity of S is complexed, since $[E]_0 \ll [S]_0$. And thus $[S] \simeq [S]_0$.) The total concentration of E, E_0, is

$$[E]_0 = [E] + [E \cdot S] \tag{4-91}$$

Substituting the steady-state expression for [E·S] into Eq. (4-91) gives

$$[E]_0 = [E] + \frac{k_1 [E][S]}{k_{-1} + k_2} \tag{4-92}$$

Solving for [E] and substituting into Eq. (4-89) gives

$$\frac{-d[S]}{dt} = \frac{d[P]}{dt} = \frac{k_1 k_2 [S]}{k_{-1} + k_2 + k_1 [S]} [E]_0 \tag{4-93}$$

Comparison of Eq. (4-93), known as the Michaelis-Menten equation, with the corresponding relation of Eq. (4-89) for the noncatalytic reaction immediately reveals that the difference lies in the additional $k_1 [S]$ term in the denominator. This term arises because of the concentration relations based on the catalytic nature of the reaction. It gives rise to the characteristic feature of enzyme kinetics: that the rate attains a maximum or limiting value at high [S]. The same mathematical treatment and result applies generally to catalytic systems.

The Michaelis-Menten equation can be recast in the form

$$\frac{d[P]}{dt} = \frac{k_2 [E]_0 [S]}{K_S + [S]} \tag{4-94}$$

where $K_S = (k_{-1} + k_2)/k_1$ is known as the *Michaelis constant*. The reader can easily show that K_S represents the concentration of substrate at which the rate is half its

---

[†] The original stoichiometry S + E = P becomes S = P. We are assuming the reaction is irreversible ($k_{-2} \cong 0$) or is being studied by initial rate methods permitting the neglect of the back reaction.

maximum value, the latter being $k_2[E]_0$. The parameters K_S and k_2 can be evaluated by a number of standard methods (see Prob. 4-10). The quantity k_2 (or maximum rate/$[E]_0$) is also known as the *turnover number*. It is a direct measure of the catalytic efficiency of the enzyme, and it represents the maximum number of moles of product formed per unit time.

The simple Michaelis-Menten mechanism just considered is limited in these respects: (1) it ignores the back reaction; (2) it allows only an enzyme-substrate complex; and (3) it does not allow for the possibility of enzyme-product interaction. A more complete mechanism is the following:

$$E + S \underset{-1}{\overset{1}{\rightleftharpoons}} E{\cdot}S \underset{-2}{\overset{2}{\rightleftharpoons}} E{\cdot}P \underset{-3}{\overset{3}{\rightleftharpoons}} E + P \tag{4-95}$$

In this case, however, there are a maximum of four experimentally determined parameters, the Michaelis constants for forward and reverse processes, K_S and K_P (with $K_P = (k_{-1}k_{-2} + k_{-1}k_{-3} + k_2k_3)/k_{-3}(k_{-2} + k_2 + k_{-1}))$, and the maximum rate in each direction.

Certain substances known as *competitive inhibitors* may lower the catalytic efficiency of the enzyme by binding reversibly to it ($E + I \rightleftharpoons E{\cdot}I$, with formation constant K_I). Other modes of inhibition are also possible, as the inhibitor may instead bind to the intermediate enzyme-substrate complex, forming $E{\cdot}S{\cdot}I$, or to a neighboring site on the enzyme but not the active site. In the case of competitive inhibition, the rate law is (see Prob. 4-11)

$$\frac{-d[S]}{dt} = \frac{k_2[E]_0[S]}{k_1[S] + K_S(1 + [I]K_I^{-1})} \tag{4-96}$$

Studies of the rate as a function of $[I]$ can be used to evaluate the binding constant K_I, and the effectiveness of different inhibitors having the same functional groups in similar structures can be used to infer the specificity of the enzymatic catalyst and perhaps the nature of its active site.

A quite analogous treatment may be applied to the kinetics of heterogeneous catalysis. Consider a surface S that contains s_0 active sites to which reactant A, a gaseous or liquid solute species, may bind reversibly:

$$A + S \overset{K_A}{\rightleftharpoons} A{\cdot}S \tag{4-97}$$

Regardless of subsequent processes, the fraction of surface sites occupied by A, Θ_A (known as the *Langmuir adsorption isotherm*), is

$$\Theta_A = \frac{[A{\cdot}S]}{s_0} \tag{4-98}$$

$$\Theta_A = \frac{K_A[A]}{1 + K_A[A]} \tag{4-99}$$

The subsequent chemistry may consist of steps such as the following:

$$A{\cdot}S \overset{k_1}{\longrightarrow} \text{product} \tag{4-100a}$$

$$B + A \cdot S \xrightarrow{k_2} \text{products} \tag{4-100b}$$

$$B + S \underset{}{\overset{K_B}{\rightleftharpoons}} B \cdot S \tag{4-100c}$$

$$A \cdot S + B \cdot S \xrightarrow{k_3} \text{products} \tag{4-100d}$$

where, if (4-100c) occurs—a process analogous to competitive inhibition if B·S is inactive—Θ_A becomes $K_A[A](1 + K_A[A] + K_B[B])$. For the last possibility, Eq. (4-100d), the rate law is

$$\frac{-d[A]}{dt} = \frac{k_3 K_A K_B s_0 [A][B]}{(1 + K_A[A] + K_B[B])^2} \tag{4-101}$$

which can clearly attain any number of limiting forms depending on the concentrations of A and B and their respective binding constants. In these mechanisms product binding has been assumed to be negligible. If that is not the case, additional denominator terms will appear.

It should also be noted that this treatment has assumed that diffusion of solute to the surface is not rate-limiting (generally true) and that all surface sites are equivalent. The latter is often not the case, and different surface planes of single metal crystals have been shown to have markedly different activity.[14]

4-9 NUMERICAL SOLUTIONS TO RATE EQUATIONS

Reaction mechanisms of some complexity often pose special problems. The coupled differential equations may not have closed-form solutions; or if such solutions do exist, they may be too complex for use in the kinetic treatment at hand. Such situations may occur when a reaction intermediate fails to satisfy the steady-state approximation, when several reactions occur in sequence, or when the mechanism consists of any complex array of parallel, competitive, reversible, and sequential reactions.

There are two parts to the problem: (1) the generation of the concentration of each substance at time intervals throughout the course of the run, given values of all the pertinent rate constants, and (2) refinement of the rate constant data to optimize the fit of the observed and calculated concentration-time or property-time data.

Two methods address the first point. They are the use of analog computers and the application of techniques of numerical integration, often via the fourth-order Runge-Kutta method. Analog computers became neglected with the advent of high-speed digital computers. Their actual operation is quite simple, although the principles are too complex to enumerate here.[15] The output consists of a plot of each concentration, or of a composite property, against time. It is usually an easy matter to vary one resistor (one rate constant) at a time to see its effect on the kinetic curves.

The fourth-order Runge-Kutta method[16] is a relatively simple procedure for numerical integration which approximates each concentration derivative dC_i/dt as $\Delta C_i/\Delta t$ in sufficiently small increments that the error is negligible. The procedure can be carried out quite easily by a computer or even a small electronic calculator, and it produces a table giving each concentration at every desired time. The frequency of the time interval can be made as small as desired to attain the accuracy needed. The background and simple applications have been described,[17] as have been the computer programs[18] needed to treat more complex schemes.

With the Runge-Kutta method one is left to compare the output with the experimental results, but the method provides no means of varying the rate constants to seek the optimum solution. Trial-and-error solutions are tedious and run the risk of false minima, particularly in complex situations. The "method of steepest descent" described by Wiberg[17] can be used, and least-squares optimization has been successful in some instances.[18,19]

REFERENCES

1. Benson, S. W.: "The Foundations of Chemical Kinetics," McGraw-Hill Book Company, New York, 1969, pp. 39–42.
2. Szabo, Z. G.: in C. H. Bamford and C. F. H. Tipper (eds.), "Comprehensive Chemical Kinetics," Elsevier Scientific Publishing Co., Inc., New York, vol. 2, pp. 20–24.
3. Rodiguin, N. M., and E. N. Rodiguina: "Consecutive Chemical Reactions," English ed., translated by R. F. Schneider, D. Van Nostrand Company, Princeton, N.J., 1964.
4. Alcock, N. W., D. J. Denton, and P. Moore: *Trans. Faraday Soc.*, **66**:2210 (1970).
5. Deutsch, C.: in D. F. DeTar (ed.), "Computer Programs for Chemistry," vol. IV, Academic Press, New York, 1972.
6. Buckingham, D. A., D. J. Francis, and A. M. Sargeson: *Inorg. Chem.*, **13**:2630 (1974).
7. Wilkins, R. G.: "The Study of Kinetics and Mechanisms of Reactions of Transition Metal Complexes," Allyn & Bacon, Inc., Boston, 1974, pp. 23–25, and references therein.
8. Chipperfield, J. R.: *J. Organometal. Chem.*, **137**:355 (1977).
9. (*a*) Vanderheiden, D. R., and E. L. King: *J. Am. Chem. Soc.*, **95**:3860 (1973); (*b*) W. Marty and J. H. Espenson: *Inorg. Chem.*, **18**:1246 (1979).
10. Fersht, A. R., and W. P. Jencks: *J. Am. Chem. Soc.*, **92**:5432 (1970).
11. Johnston, H. S.: "Gas Phase Reaction Rate Theory," Ronald Press, New York, 1969, pp. 329–332.
12. Volk, L., W. Richardson, K. H. Lau, M. Hall, and S. H. Lin: *J. Chem., Educ.*, **54**:95 (1977).
13. Fawcett, J. P., A. Poë, and K. R. Sharma: *J. Am. Chem., Soc.*, **98**:1401 (1976).
14. McAllister, J., and R. S. Hansen: *J. Chem. Phys.*, **59**:414 (1973).
15. Crossley, T. R., and M. A. Slifkin: *Progr. Reaction Kinetics*, **5**:409 (1970).
16. Margenau, H., and G. M. Murphy: "The Mathematics of Physics and Chemistry," D. Van Nostrand Company, New York, 1943, p. 469.
17. Wiberg, K.: in E. S. Lewis (ed.), "Investigations of Rates and Mechanisms of Reactions: General Considerations and Reactions at Conventional Rates," vol. VI, part 1 of the series "Techniques of Chemistry," Wiley-Interscience, New York, 1974, pp. 764–769.
18. Newton, T. W., and F. B. Baker: *J. Phys. Chem.*, **70**:1943 (1968).
19. Mønsted, L., and O. Mønsted: *Acta Chem. Scand.*, **A32**:19 (1978).

PROBLEMS

4-1 *Consecutive reactions.* Consider the reaction scheme of Eq. (4-1).

(*a*) Sketch on a single graph [A], [B], and [C] versus time. The plot should show the correct qualitative features for the case $k_1 \gg k_2$. Likewise for $k_1 \ll k_2$.

(b) If $[B]_0 \neq 0$ in this scheme, prove that $[B]$ will go through a maximum only under the condition $k_1[A]_0 > k_2[B]_0$.

4-2 *Consecutive reactions.* Consider the reaction scheme of Eq. (4-1) when both steps have identical rate constants, $k_1 = k_2 = k$.

(a) Solve the differential equations and present solutions for $[A]$, $[B]$, and $[C]$ with the boundary condition $[B]_0 = [C]_0 = 0$.

(b) Derive expressions for $[B]_{max}$ and t^B_{max} in terms of k and/or $[A]_0$.

(c) Sketch the concentrations of all three substances versus time on a single graph. Show accurately the important features such as location of maxima and points of inflection.

4-3 *Consecutive reactions.* Derive the condition stated by Eq. (4-19). Simplify the resulting kinetic expression obtained when the result is substituted into Eq. (4-12).

4-4 *Steady-state approximation.* H. C. Brown and C. J. Kim [*J. Am. Chem. Soc.*, **93**:5765 (1971)] have done kinetic studies on the solvolysis of *sec-β-aryl-alkyl* tosylates, and they suggest the following mechanism:

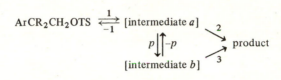

Prove the following expression for the rate constant for product formation, making the steady-state assumption for a and b, where $F = k_3/(k_3 + k_{-p})$.

$$k = \frac{k_1(k_2 + Fk_p)}{k_{-1} + k_2 + Fk_p}$$

4-5 *Steady-state approximation; successive steps.* One leading textbook has given an erroneous expression for the rate of reaction of Fe^{3+} and I^-. The mechanism and rate law are shown below; derive a correct expression making the steady-state assumption for $[I_2^-]$ and $[FeI^{2+}]$

$$I^- + Fe^{3+} \underset{-1}{\overset{1}{\rightleftharpoons}} FeI^{2+}$$

$$I^- + FeI^{2+} \underset{-2}{\overset{2}{\rightleftharpoons}} Fe^{2+} + I_2^-$$

$$I_2^- + Fe^{3+} \overset{3}{\longrightarrow} I_2 + Fe^{2+}$$

$$\frac{-d[Fe^{3+}]}{dt} = \frac{2k_1k_2k_3[I^-]^2[Fe^{3+}]}{\dfrac{k_{-1}k_{-2}[Fe^{2+}]}{[Fe^{3+}]} + k_{-1}k_3 + \dfrac{k_2k_{-2}[Fe^{2+}][I^-]}{[Fe^{3+}]} + k_2k_3[I^-]}$$

4-6 *Steady-state approximation; parallel pathways.* The coupling of *o*-bromonitrobenzene ($=ArBr$) induced by Cu(I) has been studied recently [T. Cohen and I. Cristea, *J. Am. Chem. Soc.*, **98**:748 (1976)]. They find mainly coupling product ($=Ar_2$) accompanied by a smaller amount of nitrobenzene ($=ArH$) by-product, and they formulate a separate rate law for each product by using the initial rate of product production at various reactant concentrations. They report these rate laws

$$\frac{d[Ar_2]}{dt} = k_{Ar_2}[ArBr]^2[Cu(I)]$$

$$\frac{d[ArH]}{dt} = k_H[ArBr][Cu(I)]$$

with $k_{Ar_2} = 0.16 \; M^{-2} \; s^{-1}$ and $k_H = 1.8 \times 10^{-5} \; M^{-1} \; s^{-1}$. They suggest the following mechanism

$$ArBr + Cu(I) \underset{k_{-1}}{\overset{k_1}{\rightleftarrows}} ArCu^{III}Br$$

$$ArCu^{III}Br + ArBr \xrightarrow{k_2} Ar_2 + Cu^{III}Br_2$$

$$ArCu^{III}Br \xrightarrow{k_3} ArH + Cu^{III}Br$$

(a) Apply the steady-state approximation to the intermediate $ArCu^{III}Br$, and derive an expression for the rate of formation of each product.

(b) What inequality must apply in each case to reduce the answers in part a to the experimental form?

(c) Compute a value for k_2/k_3.

4-7 *Steady-state approximation; competition experiments.* R. A. Sneen and J. W. Larsen [*J. Amer. Chem. Soc.*, **91**:362 (1969)] have studied the solvolysis of 2-octylmethanesulfonate (RX) in water-azide mixtures where it is converted to a mixture of the alcohol (ROH) and the azide (RN$_3$). They consider two mechanisms shown below.

Scheme I:

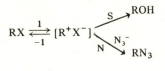

Scheme II:

For each mechanism (a) derive an expression for the pseudo-first-order rate constant for the disappearance of RX and (b) derive an expression for the yield [RN$_3$]/[ROH] in terms of rate constants and [N$_3^-$]. (c) Use the following data (in 25% aqueous dioxane at 36.2°C) to distinguish between the alternatives. Present graphs or computations to illustrate the point convincingly. For the correct mechanism give values for the rate constants or rate constant ratios which can be computed from these data.

[N$_3^-$]	Rate constant,[a] k_{obs}/s^{-1}	Product ratio, [RN$_3$]/[ROH]
0.00	2.21×10^{-4}	0.00
0.076	3.02×10^{-4}	0.625
0.156	3.64×10^{-4}	1.28
0.237	4.11×10^{-4}	1.95

[a]Equals $-d \ln [RX]/dt$ under pseudo-first-order conditions.

4-8 *Steady-state approximation; product isomer ratios.* The cycloaddition reaction of 1-benzyne and *cis*-1,2 dichloroethylene [Jones, M., Jr. and R. H. Levin: *J. Am. Chem. Soc.*, **91**:6411 (1969)] proceeds to a mixture of cis and trans products according to the mechanism (the 2 diradicals are steady-state intermediates):

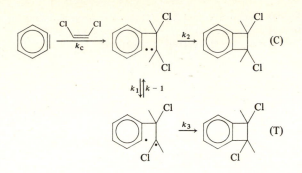

(a) Prove that the ratio of cis and trans products is given by the expression

$$\frac{[C]}{[T]} = \frac{k_2(k_{-1} + k_3)}{k_1 k_3} = R_{cis}$$

(b) Assuming the identical mechanism for the reactions of 1-benzyne with *trans*-1,2-$C_2H_2Cl_2$, except that the lower intermediate is formed first, i.e.,

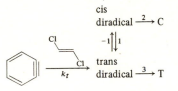

derive an expression analogous to that in (a) for the yield of cis/trans. (The yield of [C]/[T] in an experiment where trans is the starting material is designated R_{trans}.)

(c) Compute the ratios k_1/k_2 and k_3/k_{-1} from these ratios, $R_{cis} = 2.12$ and $R_{trans} = 0.230$.

4-9 *Steady-state approximation; competition experiments.* R. A. Sneen and H. M. Robbins [*J. Am. Chem. Soc.*, **91**:3100 (1969)] have studied the solvolysis of α-phenylethylbromide (R-X) in ethanol-ethoxide solutions, which produces the ether (the substitution product) and styrene (the elimination product).

They suggest the following mechanism, where R^+X^- is a steady-state intermediate:

$$
\begin{array}{c}
CH_3 \\
| \\
PhCHBr\,(RX) \underset{-1}{\overset{1}{\rightleftarrows}} R^+X^-
\end{array}
\quad
\begin{array}{l}
\overset{k_{1s}}{\longrightarrow} \begin{array}{c} CH_3 \\ | \\ PhCHOEt\,(ROEt) \end{array} \\
\overset{k_{1e}}{\longrightarrow} PhCH=CH_2 \\
\overset{k_{2s}[OEt^-]}{\longrightarrow} ROEt \\
\overset{k_{2e}[OEt^-]}{\longrightarrow} PhCH=CH_2
\end{array}
$$

(a) Derive their expression for the pseudo-first-order rate constant:

$$k_{obs} = \frac{k_1\{k_{1s} + k_{1e} + (k_{2s} + k_{2e})[OEt^-]\}}{k_{-1} + k_{1s} + k_{1e} + (k_{2s} + k_{2e})[OEt^-]}$$

(b) Derive an expression for the ratio R of styrene to ether (as a function of rate constants and [OEt$^-$]).

(c) Given the following values of R and k_{obs}, compute numerical values for the following rate constant combinations: (1) k_{1e}/k_{1s}, (2) k_{2e}/k_{2s}, (3) k_{2s}/k_{1s}, (4) k_1, (5) k_{-1}/k_{1s}.

Hint: limits of R and of k_{obs} at 0 and ∞ [EtO$^-$] might help.

[EtO$^-$]	R	k_{obs}, s^{-1}
0	0.018	1.13×10^{-4}
0.20	0.362	3.8×10^{-4}
0.60	0.443	6.2×10^{-4}
1.00	0.464	7.4×10^{-4}
∞	0.500	10.6×10^{-4} (extrapolated)

4-10 *Enzyme kinetics.* Kinetic data for reactions following the Michaelis-Menten mechanism are frequently analyzed by a plot (known as the Hanes plot) of $1/k_{obs}$ ($k_{obs} = -d \ln [S]/dt$) against [S]. Other authors choose to make a plot of $(-d[S]/dt)^{-1}$ versus [S]$^{-1}$; this double-reciprocal plot, known as the Lineweaver-Burk plot, requires a value of the initial reaction rate itself. Show that each plot is expected to be linear, and in each case give the relations needed to calculate k_2 and the Michaelis constant K_S from the slope and intercept of the line.

4-11 *Competitive inhibition.* Derive Eq. (4-96), applicable to the mechanism (for the case $[E]_0 \ll [S]$ and [I]).

$$E + S \underset{-1}{\overset{1}{\rightleftharpoons}} E \cdot S$$

$$E + I \overset{K_I}{\rightleftharpoons} E \cdot I$$

$$E \cdot S \overset{k_2}{\longrightarrow} E + P$$

DEDUCTION OF REACTION MECHANISMS

The route leading from kinetic data to reaction mechanism is not always a straight-forward one. The first step is to convert the experimental data, which most often take the form of concentration as a function of time, to a differential rate expression, which gives reaction rate as a function of concentration(s). The first four chapters of this book have been concerned with that aspect of the problem. The next consideration is the interpretation of the rate law to reveal the family of chemical reactions which comprise the mechanism.

Once one mechanism which is consistent with the data at hand is found, two further considerations immediately arise. First, what other mechanisms will also fit the existing data? Second, are there additional experiments, kinetic determinations or otherwise, which permit some of the possibilities to be ruled out?

5-1 THE ACTIVATED COMPLEX OR TRANSITION STATE

Every elementary reaction proceeds through a critical complex corresponding to the state of potential-energy maximum along its reaction coordinate. In kinetics we are primarily concerned with the highest energy step (or steps) along the entire potential-energy surface, which corresponds to the critical activation process. To a lesser extent we may also consider any rapid reactions which precede or follow the rate-limiting step(s).

One goal of both theory and experimental kinetics is a full understanding of the activation process—the evolution of structure and energetics accompanying the molecular transformations in the reactants as they proceed along the reaction coordinate toward the activated complex. This level of understanding has as yet been attained only in some quite simple elementary reactions.

For most "real" systems we settle for far less, and we will seek the answers to questions such as the number of transition states, whether they occur sequentially or in parallel, and whether chemical reaction intermediates are involved. Kinetic data and related experiments will answer those questions, particularly those relating to the composition of the activated complex.

We need to distinguish between an *intermediate* and an *activated complex*. The former lies at a potential-energy minimum along the reaction coordinate: Further activation, which might be intramolecular distortion or rearrangement, or subsequent bimolecular reaction with a different chemical component, is needed to convert the intermediate to the activated complex leading to products. The activated complex is necessarily decomposing into products of reaction (or to substances which themselves lead to the products); that is the only type of reaction it can undergo. The intermediate was itself preceded by another activated complex; both activated complexes lie at potential-energy maxima.

An intermediate might coincidentally have the same composition as an activated complex, but it is not identical with it. The intermediate has a choice of reactions; among the reactions are reconversion to starting material and reaction with a trapping reagent. It might have nearly the same geometry as the activated complex as well, but some distortion or other activation process would be needed before the intermediate could become an activated complex.

The concentrations appearing in the rate expression provide directly the "composition" (the elements which are present and the ionic charge, if any) of each activated complex for every elementary reaction in the mechanistic array that can be a rate-limiting step under the conditions studied. It therefore gives the number of separate activation steps needed, and it provides an indication of whether they occur in parallel or sequentially.

5-2 MECHANISTIC INTERPRETATION OF RATE LAWS

Certain features of mechanism manifest themselves in a regular way, and they serve as guides in formulating one or more mechanisms consistent with the data. Listed here is this set of "rules," more properly to be regarded as "clues," useful in formulating the mechanism. Later sections in this chapter amplify certain points.

1. *The concentration dependences in the rate law establish the elemental composition and charge of the activated complex(es).*

The rate law for the reaction of ClO_2^- and I^- in acidic solution is

$$\frac{-d[ClO_2^-]}{dt} = k[ClO_2^-][I^-][H^+] \tag{5-1}$$

This rate law is consistent with a single activated complex of composition $[HClO_2I^- \pm n(H_2O)]^\ddagger$. The involvement of solvent water cannot be defined, since its concentration is not varied. A second example is the oxidation of oxalic acid by chlorine,

$$Cl_2 + (COOH)_2 = 2CO_2 + 2Cl^- + 2H^+$$

which follows the rate law

$$\frac{d[CO_2]}{dt} = \frac{k[Cl_2][(COOH)_2]}{[H^+]^2[Cl^-]} \tag{5-2}$$

The composition of the activated complex is thus $[Cl_2 + (COOH)_2 - 2H^+ - Cl^- \pm nH_2O]^\ddagger$, or $[C_2O_4Cl(H_2O)_n^-]^\ddagger$. If the rate law consists of more than a single term, either numerator or denominator terms, the mechanism consists of as many activated complexes as there are terms in the rate law. The composition of each is given by the same procedures as shown here, letting each term predominate in turn. Examples are given under rules 3 and 4.

2. *A rate law to be properly interpreted according to rule 1 above must be written in terms of the predominant species in the reaction medium.*

   That is to say, we must take due account of pertinent equilibria and use the equilibrium relations to convert the experimental rate law, which will often be expressed in terms of stoichiometric concentrations, to a form containing only the concentrations of the predominant species. This point will be amplified in Sec. 5-6, and we need note here only one trivial example. The rate law of Eq. (5-1) contains the product of concentrations, $[H^+][I^-]$. Were the experiments carried out by using HI as the source of H^+ and I^- ions, the rate would be proportional to the square of the laboratory concentration, C_{HI}^2. That relation holds because HI is a strong electrolyte in water, so that $[H^+] = [I^-] = C_{HI}$. Yet were the observation taken literally, without recognition of the dissociation of HI, application of rule 1 would suggest that the activated complex contains two molecules of HI rather than the elements of one HI molecule.

3. *The number of (positive) numerator terms in a rate law is the number of parallel pathways involving transition states of different composition.*

   By way of example, the reaction $3I^- + H_2O_2 + 2H^+ = I_3^- + 2H_2O$ proceeds according to a rate law having two (numerator) terms:

$$\frac{d[I_3^-]}{dt} = k[H_2O_2][I^-] + k'[H_2O_2][H^+][I^-] \tag{5-3}$$

indicating two parallel pathways with transition states of composition $[H_2O_2I^-]^\ddagger$ and $[H_3O_2I]^\ddagger$.

   A negative numerator term results from a reverse reaction, one term for each pathway, reflecting the approach to equilibrium. The classic S_N2 mechanism for

$$RX + Y^- \underset{k_r}{\overset{k_f}{\rightleftarrows}} RY + X^-$$

has a rate law

$$\frac{-d[RX]}{dt} = k_f[Y^-][RX] - k_r[RY][X^-]$$

If the mechanism is expressed as the rate constant for the approach to equilibrium, a summation of two terms is found, which, when X^- and Y^- are present in large excess, takes the form $k_{obs} = k_f[X^-] + k_r[Y^-]$. But one is careful to note that this corresponds to one activated complex, not two. That is, the summation of terms arises in this case because the expression for k_{obs} is not the expression for the reaction rate itself.

4. *A summation of terms in the denominator indicates a mechanism consisting of successive steps; one or more of the steps is a reversible reaction.*

   The number of denominator terms indicates the number of sequential activated complexes. That is, it gives the number of reaction steps, excluding, of course, any rapid reactions occurring before or after. The denominator terms represent the competitive reactions of an intermediate, and the inequalities of the rate constant–concentration composites which lead to a given limiting form are equivalent to a change in rate-limiting step from one of the reaction steps to the other. The composition of each transition state in the succession of reaction steps is inferred according to rule 1.
   The reaction $2Fe^{2+} + Tl^{3+} = 2Fe^{3+} + Tl^+$ constitutes an instructive example. It follows the rate law

$$\frac{-d[Tl^{3+}]}{dt} = \frac{k[Fe^{2+}]^2[Tl^{3+}]}{[Fe^{2+}] + k'[Fe^{3+}]} \tag{5-4}$$

This requires two activated complexes whose compositions are obtained by considering first one denominator term and then the other to predominate. The limiting rate equation when $[Fe^{2+}] \gg k'[Fe^{3+}]$ is $-d[Tl^{3+}]/dt = k[Fe^{2+}][Tl^{3+}]$, with activated complex $[TlFe^{5+}]^{\ddagger}$. With the reverse inequality, the rate is proportional to $[Fe^{2+}]^2[Tl^{3+}]/[Fe^{3+}]$, which corresponds to the activated complex $[TlFe^{4+}]^{\ddagger}$. A mechanism consistent with that is the following, where Tl^{2+} is an intermediate for which the steady-state approximation is valid:

$$Fe^{2+} + Tl^{3+} \;\rightleftharpoons\; Fe^{3+} + Tl^{2+} \tag{5-5}$$

$$Fe^{2+} + Tl^{2+} \longrightarrow Fe^{3+} + Tl^+ \tag{5-6}$$

5. *Concentrations appearing in single-term denominators are the chemical products of step(s) prior to the rate-limiting step.*

   This is a special case of rule 4. The rate law for chlorine and oxalic acid, Eq. (5-2), suggests that two H^+ ions and one Cl^- ion are produced in equilibria established rapidly (compared to the overall rate). Such equilibria *may be* (see Sec. 5-3) the following:

$$Cl_2 + H_2O \rightleftharpoons HOCl + Cl^- + H^+ \tag{5-7}$$

$$H_2C_2O_4 \rightleftharpoons H^+ + HC_2O_4^-$$ (5-8)

Needless to say, we do not learn any details about the *mechanisms* of such rapid equilibria.

6. *The sum of the mechanistic steps must be the overall chemical reaction; rapid reactions may follow the rate-limiting step(s).*

Thus the reaction $2Cr^{2+} + Tl^{3+} = 2Cr^{3+} + Tl^+$ occurs by the rate law

$$\frac{-d[Tl^{3+}]}{dt} = k[Cr^{2+}][Tl^{3+}]$$ (5-9)

consistent with either of these mechanisms:

Mechanism I: $Cr^{2+} + Tl^{3+} \longrightarrow Cr^{3+} + Tl^{2+}$ slow   (5-10)

$Cr^{2+} + Tl^{2+} \longrightarrow Cr^{3+} + Tl^+$ rapid   (5-11)

Mechanism II: $Cr^{2+} + Tl^{3+} \longrightarrow Cr^{4+} + Tl^+$ slow   (5-12)

$Cr^{2+} + Cr^{4+} \longrightarrow 2Cr^{3+}$ rapid   (5-13)

where the alternative rapid reactions are added for sake of consistency with the overall stoichiometry of the reaction. The following points deserve note: (*a*) the information in Eq. (5-9) does not distinguish between the alternatives; (*b*) this kinetic study provides no information concerning the mechanism of either postulated rapid reaction, (5-11) or (5-13); (*c*) the lower power for $[Cr^{2+}]$ in the kinetic equation compared to the stoichiometric reaction (1 versus 2) indicates that a reaction intermediate is produced and then consumed in a subsequent rapid step.

7. *Alternative mechanisms leading to the same pattern of activated complexes are not kinetically distinguishable.*

Since both steps 10 and 12 have an activated complex $[CrTl^{5+}]^{\ddagger}$, the rate law of Eq. (5-9) cannot afford a choice between them. Among the further options we might consider are these two: (*a*) a wider range of concentration variation, particularly experiments with added Cr^{3+}, added Tl^+, and/or a lower concentration of Cr^{2+}, may elicit additional kinetic terms and transform Eq. (5-9) to a form analogous to the more informative Eq. (5-4). (*b*) A comparison of the chemistry of the proposed intermediates Tl^{2+} and Cr^{4+} to suggest which is the more reasonable formulation. The evidence may be thermodynamic (dealing with the stability of such intermediates), stereochemical, structural, or chemical or it may be based upon the reaction products formed. Here the experimenter has the chance to bring skill and ingenuity to bear on the problem of providing critical tests of mechanism. In the present instance the $Cr(III)$ product proved to be $[(H_2O)_4Cr(OH)_2Cr(OH_2)_4]^{4+}$ rather than $Cr(H_2O)_6^{3+}$, which is suggestive of mechanism II.

8. *An increase of reaction order with increasing concentration suggests parallel pathways, whereas a decrease in order suggests a sequence of steps.*

This is an alternative expression of statements 3 and 4. Note the order with respect to $[H^+]$ increases with $[H^+]$ in Eq. (5-3), whereas the orders in $[Fe^{2+}]$ and $[Fe^{3+}]$ in reaction (5-4) decrease with an increase in the concentration of the ion in question.

5-3 EQUIVALENT KINETIC EXPRESSIONS

As stated earlier, the rate of reaction of oxalic acid and chlorine,

$$Cl_2 + H_2C_2O_4 = 2CO_2 + 2Cl^- + 2H^+ \tag{5-14}$$

when expressed in terms of the predominant species in solutions, is given by[1]

$$\frac{-d[Cl_2]}{dt} = k \frac{[Cl_2][(COOH)_2]}{[H^+]^2[Cl^-]} \tag{5-15}$$

The following are rapidly established equilibria:

$$Cl_2 + H_2O \overset{K_d}{\rightleftharpoons} HOCl + Cl^- + H^+ \tag{5-16}$$

$$HOCl \overset{K_a}{\rightleftharpoons} H^+ + OCl^- \tag{5-17}$$

$$H_2C_2O_4 \overset{K_1}{\rightleftharpoons} H^+ + HC_2O_4^- \tag{5-18}$$

$$HC_2O_4^- \overset{K_2}{\rightleftharpoons} H^+ + C_2O_4^{2-} \tag{5-19}$$

By appropriate substitution, it can be readily shown that the following expressions are equivalent to the original rate expression

$$k_1 \frac{[Cl_2][HC_2O_4^-]}{[H^+][Cl^-]} \qquad k_2 \frac{[Cl_2][C_2O_4^{2-}]}{[Cl^-]} \tag{5-20}$$

$$k_3[HOCl][HC_2O_4^-] \qquad k_4[OCl^-][H_2C_2O_4]$$

It is pointless to argue whether the kinetic data support a rate-limiting step consisting of the bimolecular reaction of HOCl with $HC_2O_4^-$ or of OCl^- with $H_2C_2O_4$. The rate law prescribes *only* the composition of the activated complex—and neither the order in which the components come together nor the mode by which they are bonded. Perhaps "chemical" arguments or comparison reactions[†] can make a case for one formulation or the other, but the bare kinetic equations do not.

It might be commented that Eq. (5-15) remains the best form in which to express the data obtained in acidic solution, because the rate is expressed in terms of the species predominating under those circumstances. If one then wishes to interpret the data in terms of a specific bimolecular process, say the reaction of $HC_2O_4^-$ with HOCl,

---

[†]For example: Does the dimethyl ester of oxalic acid react with OCl^-? Does the monomethyl ester react with HOCl?

then a derived rate constant [k_3 of Eq. (5-20)] can be calculated from the relation

$$k_3 = \frac{k}{K_d K_1} \tag{5-21}$$

This point is especially well kept in mind if the equilibrium constants are known only very poorly, as may be the case for very feeble associations, or if a long extrapolation of experimental conditions is required. In such cases, one should always report the rate constant values resulting directly from the experimental conditions. For purposes of discussion or comparison, one might later wish to convert the precise experimental value into an approximate derived value, recognizing all the while that the latter constitutes an interpretation of the investigator. In so doing it is assumed first that the choice of elementary mechanistic steps is correct and secondly that the necessary equilibrium constant(s) can be estimated with sufficient precision for the comparison purposes at hand.

By way of example, Hicks and Sutter[2] found that the reaction of permanganate ion with tris-(1,10-phenanthroline)iron(II) ion contained a kinetic term of the form

$$\frac{-d[MnO_4^-]}{dt} = (k_1 + k_2[H^+])[MnO_4^-][Fe(phen)_3^{2+}] \tag{5-22}$$

and gave these values: $k_1 = (6.1 \pm 2.2) \times 10^3$ M^{-1} s^{-1}, $k_2 = (7.42 \pm 0.11) \times 10^5$ M^{-2} s^{-1}. If the second term is *interpreted* as being the bimolecular reaction of permanganic acid with $Fe(phen)_3^{2+}$,

$$HMnO_4 + Fe(phen)_3^{2+} \xrightarrow{k_2'} HMnO_4^- + Fe(phen)_3^{2+}$$

then the bimolecular rate constant k_2' can be calculated from the experimental k_2 and K_a for permanganic acid.

$$HMnO_4 \underset{K_a}{\rightleftharpoons} H^+ + MnO_4^- \tag{5-23}$$

The value for K_a is $\sim 3 \times 10^2$ M, which gives $k_2' = k_2 K_a = \sim 2 \times 10^8$ M^{-1} s^{-1}. Clearly this value of K_a is *highly* approximate, considering the practical difficulties attending the evaluation of the ionization constant of such a strong acid. The approximate k_2' may nonetheless be useful in that it can be directly compared with k_1; for example, one might note that $HMnO_4$ is some $10^{4.7}$ times more reactive than MnO_4^- toward $Fe(phen)_3^{2+}$. The value of the direct experimental value of k_2 is, however, likely to be of more lasting significance than the derived and approximate k_2'. Workers finding a situation such as that should be certain to report the former, whether or not they choose to estimate the latter for comparison purposes. It is also to be noted that the assignment of Eq. (5-23) as the mechanism for this pathway is only one of the possible interpretations; protonation of the other reactant followed by the bimolecular reaction of MnO_4^- and $Fe(phen)_2(phenH)^{3+}$ is a realistic alternative.

Another example is afforded by the reaction of iodide and hypochlorite ions,

$$I^- + OCl^- = OI^- + Cl^- \tag{5-24}$$

in 0.1 to 1.0 M potassium hydroxide. The rate expression is

$$\frac{d[OI^-]}{dt} = \frac{k[I^-][OCl^-]}{[OH^-]} \tag{5-25}$$

Given the uncertainty in the order with respect to solvent, the composition of the activated complex is $[H_2O + I^- + OCl^- - OH^-]^{\ddagger}$. This suggests the following mechanism:

$$OCl^- + H_2O \rightleftharpoons HOCl + OH^- \quad \text{rapid} \tag{5-26}$$

$$I^- + HOCl \longrightarrow HOI + Cl^- \quad \text{slow} \tag{5-27}$$

$$\underline{OH^- + HOI = H_2O + IO^-} \quad \text{rapid} \tag{5-28}$$

$$I^- + OCl^- = OI^- + Cl^- \quad \text{overall} \tag{5-29}$$

Thus

$$\frac{d[IO^-]}{dt} = k_{27}[I^-][HOCl] \tag{5-30}$$

$$\frac{d[IO^-]}{dt} = \frac{k_{27}K_{26}[I^-][OCl^-]}{[OH^-]} \tag{5-31}$$

Because the experiments were done in strongly basic solution, Eq. (5-25) is a better representation of the actual results than (5-30). The latter may be useful for other purposes, however, particularly for the discussion of mechanism. The elementary rate constant k_{27} may be calculated by the relation $k_{27} = k/K_{26}$, provided K_{26} has a known value.

The maximum rate of a bimolecular reaction is limited in solution by the diffusion and in the gas phase by the collision frequency of the two solutes. The theoretical considerations are given later, Secs. 9-2 and 8-1, but it is helpful to consider the implications of those results in the present discussion of mechanistic alternatives. The maximum rate constant for a bimolecular step is ca. 10^{10} dm^3 mol^{-1} s^{-1} in water at 298 K. (The value differs little with solvent and temperature, but it may be modified somewhat if the reactants are ions.) Suppose that a mechanism under consideration contains a bimolecular step having a rate constant considerably in excess of that value. The mechanism, or at least this step, can consequently be ruled out. In effect, this prohibits mechanisms requiring the reaction of a species accessible only by quite unfavorable equilibria. An illustration is helpful. The reaction of $Cr^{2+}(aq)$ and $Co(NH_3)_5H_2O^{3+}$ according to the equation

$$Cr^{2+}(aq) + Co(NH_3)_5H_2O^{3+} + 5H^+ = Cr^{3+} + Co^{2+} + H_2O + 5NH_4^+$$

follows the rate law

$$\frac{-d[Cr^{2+}]}{dt} = \frac{k[Cr^{2+}][Co(NH_3)_5H_2O^{3+}]}{[H^+]}$$

with $k = 3.1$ dm^3 mol^{-1} s^{-1}. Two ready interpretations come to mind: One metal complex undergoes hydrolysis,

$$Cr^{2+} + H_2O = CrOH^+ + H^+ \quad \text{(rapid, } K_{a,Cr} \approx 10^{-11} \text{ mol dm}^{-3})$$

$$Co(NH_3)_5H_2O^{3+} + H_2O = Co(NH_3)_5OH^{2+} + H_3O^+$$

$$\text{(rapid, } K_{a,Co} = 2.5 \times 10^{-7} \text{ mol dm}^{-3})$$

and the hydrolyzed form reacts with the other metal ion complex. The alternative rate-limiting steps are then

$$CrOH^+ + Co(NH_3)_5H_2O^{3+} \xrightarrow{k_1} CrOH^{2+} + Co(NH_3)_5H_2O^{2+}$$

$$Cr^{2+} + Co(NH_3)_5OH^{2+} \xrightarrow{k_2} Cr^{3+} + Co(NH_3)_5OH^+$$

In both cases these processes would be followed by further rapid steps which are easily written down. The values of the rate constants for the elementary reactions are then $k_1 = k/K_{a,Cr} \approx 3 \times 10^{11}$ dm^3 mol^{-1} s^{-1} and $k_2 = k/K_{a,Co} = 1.2 \times 10^7$ dm^3 mol^{-1} s^{-1}. The former can be ruled out on the basis that the value of k_1 exceeds by a factor of about 30 the maximum limit allowed; in fact, the discrepancy is undoubtedly greater than that considering the ionic charges have the same signs.

5-4 PARALLEL PATHWAYS

Consider the reaction

$$3I^- + H_2O_2 + 2H^+ = I_3^- + 2H_2O \tag{5-32}$$

which proceeds according to the rate law

$$\frac{d[I_3^-]}{dt} = k_2[H_2O_2][I^-] + k_3[H_2O_2][H^+][I^-] \tag{5-33}$$

Clearly, the reaction occurs by two independent pathways, the first of which might correspond to rate-determining S_N2 displacement of OH^- by I^-:

$$H_2O_2 + I^- \longrightarrow HOI + OH^- \quad \text{slow} \tag{5-34}$$

$$H^+ + OH^- \rightleftharpoons H_2O \quad \text{very rapid} \tag{5-35}$$

$$HOI + H^+ + 2I^- = I_3^- + H_2O \quad \text{rapid} \tag{5-36}$$

The sum of the reactions corresponds to the net Eq. (5-32), and the slow step results in the proper composition for the activated complex. The rapid steps following the rate-determining step are speculative and are added primarily for completeness rather than mechanistic detail. It is only for that reason that Eq. (5-36) contains four reactants; in no way does it suggest a four-body collision. On the contrary, no mechanistic information or detail about the fast steps is obtained from this work.

The second path can be described by either of the schemes (5-37) and (5-38) or (5-39) and (5-40) followed in each case by fast reaction such as (5-36).

$$H^+ + I^- \rightleftharpoons HI \qquad\qquad \text{rapid} \qquad\qquad (5\text{-}37)$$

$$HI + H_2O_2 \longrightarrow H_2O + HOI \qquad \text{slow} \qquad\qquad (5\text{-}38)$$

$$H_2O_2 + H^+ \rightleftharpoons H_2OOH^+ \qquad \text{rapid} \qquad\qquad (5\text{-}39)$$

$$H_2OOH^+ + I^- \longrightarrow H_2O + HOI \qquad \text{slow} \qquad (5\text{-}40)$$

In any case, the second term in the rate law gives the composition of the activated complex as $[H_2O_2HI]^{\ddagger}$. The mechanistic details are somewhat less definitive; on chemical grounds one might prefer (5-39) and (5-40) over (5-37) and (5-38) (since the oxygen atom of H_2O_2 is more basic than an I^- ion and since a ready interpretation of 40 as the S_N2 displacement of H_2O by the nucleophile I^- is at hand), but such interpretation clearly goes beyond the immediate kinetic data.

5-5 SUCCESSIVE STEPS

Consider first a two-step sequence involving a single intermediate. For the net reaction

$$A + 2B = C + 2D \qquad\qquad (5\text{-}41)$$

proceeding according to this mechanism

$$A + B \underset{2}{\overset{1}{\rightleftharpoons}} X + D \qquad\qquad (5\text{-}42)$$

$$X + B \underset{4}{\overset{3}{\rightleftharpoons}} C + D \qquad\qquad (5\text{-}43)$$

the steady-state rate law can be derived by first solving for $[X]_{ss}$ and then expressing $-d[A]/dt$ in terms of concentrations of major species. Thus

$$[X]_{ss} = \frac{k_1[A][B] + k_4[C][D]}{k_2[D] + k_3[B]} \qquad\qquad (5\text{-}44)$$

$$\frac{-d[A]}{dt} = k_1[A][B] - k_2[X][D] \qquad\qquad (5\text{-}45)$$

$$\frac{-d[A]}{dt} = \frac{k_1k_3[A][B]^2 - k_2k_4[C][D]^2}{k_3[B] + k_2[D]} \qquad\qquad (5\text{-}46)$$

The negative numerator term in Eq. (5-46) arises from the reverse reaction. It is instructive to examine the form the rate law takes if the reaction is effectively irreversible. It will be if equilibrium lies far to the right, either because the equilibrium constant K $(= k_1k_3/k_2k_4)$ is very large or because forcing concentrations of reactants are used. The requisite condition for neglecting the back reaction is $[C][D]^2/[A][B]^2 \ll K$. In that event the rate law simplifies to

$$\frac{-d[A]}{dt} = \frac{k_1k_3[A][B]^2}{k_2[D] + k_3[B]} \qquad\qquad (5\text{-}47)$$

which is equivalent to having set $k_4 = 0$ in the derivation. Note that, according to Eq. (5-47), the concentration of a product D may lower the rate of a reaction even though the reverse reaction is unimportant. This is an important point: retardation by a product does not *necessarily* mean that one is approaching equilibrium. Another very common effect of a product, as shown here, is to lower the reaction rate by diverting the intermediate back to product. The term $k_2[X][D]$ in the present example represents this contribution.

In this example the order with respect to [B] decreases from 2 to 1 with increasing [B] and the order with respect to [D] also decreases from 0 to -1 with increasing [D]. The decrease in order with increase in concentration is the sign of a stepwise mechanism, the number of steps being the number of denominator terms.

The summation of denominator terms indicates competition for an intermediate. The rate law is not uniquely consistent with the mechanism shown; both of the following schemes, for example, are consistent with the kinetic data:

II:
$$A + B \rightleftharpoons Y^- + H^+ + D \tag{5-48}$$

$$Y^- + H^+ + B \longrightarrow C + D \tag{5-49}$$

III:
$$A + B \rightleftharpoons U + V + D \tag{5-50}$$

$$U + V + B \longrightarrow C + D \tag{5-51}$$

They are kinetically indistinguishable from the original for a very good reason: the transition state for each step has the same composition in each of the three mechanisms given.

A particularly interesting aspect of sequential mechanism has been pointed out by Haim,[3] who used as an example published kinetic data[4] for the reaction

$$V^{3+} + Cr^{2+} = V^{2+} + Cr^{3+} \tag{5-52}$$

for which the rate expression is

$$\frac{-d[V^{3+}]}{dt} = \frac{q[V^{3+}][Cr^{2+}]}{r + [H^+]} \tag{5-53}$$

Two fundamentally different mechanisms[†] consistent with this expression can be formulated. Each has the correct activated complexes—namely $[CrV^{5+}]^{\ddagger}$ and $[VCrOH^{4+}]^{\ddagger}$—but differs in that the order of occurrence is reversed. The first mechanism invokes the formation of an intermediate $V(OH)Cr^{4+}$ from the hydrated cations by elimination of H^+, followed by a unimolecular reaction of the intermediate:

Mechanism I:
$$Cr^{2+} + V^{3+} + H_2O \underset{-1}{\overset{1}{\rightleftharpoons}} V(OH)Cr^{4+} + H^+ \tag{5-54}$$

$$V(OH)Cr^{4+} \overset{2}{\longrightarrow} V^{2+} + CrOH^{2+} \tag{5-55}$$

$$CrOH^{2+} + H^+ \rightleftharpoons Cr^{3+} + H_2O \tag{5-56}$$

[†]Since $r = 0.108$ M and K_a for $V(H_2O)_6^{3+}$ is $\sim 2 \times 10^{-3}$ M, one need not consider mechanisms in which the two-term denominator arises from shifts in the acid ionization equilibrium. This point is elaborated upon in Sec. 5-6.

With the steady-state approximation the rate law becomes

$$\frac{-d[V^{3+}]}{dt} = \frac{(k_1 k_2/k_{-1})[Cr^{2+}][V^{3+}]}{(k_2/k_{-1}) + [H^+]} \tag{5-57}$$

which agrees with the experimental form. Haim points out, however, that a second and fundamentally quite different process[†] is also consistent with the data. In his alternative mechanism the two activated complexes occur in the opposite order, but the same intermediate is involved. The reactions are as follows.

Mechanism II:

$$V^{3+} + H_2O \rightleftharpoons VOH^{2+} + H^+ \quad (\text{rapid}, K_a) \tag{5-58}$$

$$VOH^{2+} + Cr^{2+} \underset{-3}{\overset{3}{\rightleftharpoons}} V(OH)Cr^{4+} \tag{5-59}$$

$$V(OH)Cr^{4+} + H^+ \overset{4}{\longrightarrow} V^{2+} + Cr^{3+} + H_2O \tag{5-60}$$

which lead to the rate law

$$\frac{-d[V^{3+}]}{dt} = \frac{k_3 K_a[V^{3+}][Cr^{2+}]}{(k_{-3}/k_4) + [H^+]} \tag{5-61}$$

The chemistry of the two kinetically equivalent mechanisms is really rather different: in mechanism I the intermediate reacts unimolecularly to form products, whereas protonolysis returns it to the starting materials. Just the opposite is true in mechanism II.

To quote Haim: "Although the form of the rate law defines the composition of the activated complexes, the rate law does not specify (a) the order of formation of the activated complexes, (b) the species (reactants or intermediates) which generate the activated complexes, or (c) the decomposition products (intermediates or products) of the activated complexes."[3]

Newton and Baker[5] have suggested that a strong analogy is to be found between electrical circuits and reaction mechanisms. Birk[6] has amplified that notion. It will not be considered in great detail here, and the interested reader is referred to the indicated references. The Newton-Baker method consists of constructing an electrical circuit analogous to the mechanism; resistors correspond to activated complexes, junctions between resistors to steady-state intermediates, and terminals to reactants and products. *Any other electrical circuit which would give the same overall conductance corresponds to a kinetically equivalent mechanism.* These circuits correspond to *all* of the fundamentally different mechanisms.

The electrical circuit corresponding to mechanism I for the V^{3+}-Cr^{2+} reaction,

---

[†]We shall not consider, as fundamentally different, sets of reaction steps which differ only in the mode of assembly of the same activated complex. Thus the following is not considered "different" from Eqs. (5-58) and (5-59) in the present context.

$$Cr^{2+} + H_2O \rightleftharpoons CrOH^+ + H^+ \tag{5-58a}$$

$$CrOH^+ + V^{3+} \underset{-3'}{\overset{3'}{\rightleftharpoons}} V(OH)Cr^{4+}(aq) \tag{5-59a}$$

which consists of a sequence of two activated complexes and one intermediate, is simply two resistances in series:

$$1 \qquad 2 \tag{5-62}$$

Clearly there is one (and only one) equivalent circuit, namely:

$$2 \qquad 1 \tag{5-63}$$

These arguments can be used to show that mechanisms I and II are the complete set of "fundamentally different" mechanisms which need be considered; we need not search for further possibilities. Having written one of the mechanisms, the electrical analog method would have told us that there was one other mechanism to be considered.

In the example at hand, both of the two mechanisms seem "reasonable," although perhaps chemical arguments might be advanced for one over the other. In some similar cases the second mechanism seems far less plausible. Consider, for example, the reaction of Fe^{2+} and Tl^{3+}, mentioned in Sec. 5-2. The mechanism shown in Eqs. (5-5) and (5-6) corresponds to the same circuit as in (5-62), and we therefore need to consider the alternative sequence corresponding to Eq. (5-63), in which the activated complexes occur in the reverse order. Such a mechanism is the following:

$$2Fe^{2+} \overset{K_1}{\rightleftharpoons} Fe^+ + Fe^{3+} \qquad (\text{rapid}, K_1 \ll 1) \tag{5-64}$$

$$Fe^+ + Tl^{3+} \underset{-1}{\overset{1}{\rightleftharpoons}} Fe^{2+} + Tl^{2+} \tag{5-65}$$

$$Fe^{3+} + Tl^{2+} \overset{2}{\longrightarrow} Fe^{4+} + Tl^+ \tag{5-66}$$

$$Fe^{4+} + Fe^{2+} = 2Fe^{3+} \qquad (\text{fast}) \tag{5-67}$$

which leads to the rate expression

$$\frac{-d[Tl^{3+}]}{dt} = \frac{(k_1 k_2 K_1 / k_{-1})[Fe^{2+}]^2 [Tl^{3+}]}{[Fe^{2+}] + (k_2/k_{-1})[Fe^{3+}]} \tag{5-68}$$

The rate equation agrees with the experimental form [Eq. (5-4)]. However, the mechanism seems far less likely than that of Eq. (5-6), involving as it does two highly unusual and unstable oxidation states of iron.

5-6 PREEQUILIBRIA

We shall now examine the way in which a "balanced" or "partial" equilibrium affects the algebraic form of the rate law. We have already seen cases in which an equilibrium lay very far to one side or the other; the virtually complete ionization of HI made the substitution $C_{HI} = [H^+] = [I^-]$ (Sec. 5-2) valid, and the trivial degree of formation of VOH^{2+} [Eq. (5-58)] or Fe^+ [Eq. (5-64)] gave simple expressions in terms of the concentrations of bulk reagents:

$$[VOH^{2+}] = \frac{K_a[V^{3+}]}{[H^+]} \quad \text{and} \quad [Fe^+] = \frac{K_1[Fe^{2+}]^2}{[Fe^{3+}]} \tag{5-69}$$

But the case in which the products and reactants of a mobile equilibrium are at comparable concentrations requires a separate treatment. For one thing it leads to a summation of denominator terms which have the same form as that for a successive-step mechanism, and one must take care to avoid an incorrect assignment. For another, it is necessary to have a rate law expressed in terms of the concentration of the predominant species in solution if the methods given earlier in this chapter are to be used to deduce the mechanism. To achieve that, proper account must be taken of all equilibrium steps which proceed to an appreciable extent.

Toward that end we first introduce some notation. The symbol C_i, or sometimes $[i]_T$, will be used to represent the *stoichiometric concentration*[†] of substance i; the *molar concentration* of the actual molecular or ionic species derived from substance i and in equilibrium with it are designated by the usual square brackets, and the values add to the total. Thus a dilute, acidic solution of chromium(VI) contains the species $HCrO_4^-$, $Cr_2O_7^{2-}$, and H_2CrO_4, and the following relation applies

$$C_{Cr(VI)} = [Cr(VI)]_T = [HCrO_4^-] + 2[Cr_2O_7^{2-}] + [H_2CrO_4] \tag{5-70}$$

It is useful to account for effects of such balanced equilibria in terms of the fractional degree of conversion of one species to the other. For an equilibrium written as a dissociation, such as the ionization of the weak acid HA, the relations are

$$\alpha_{HA} = \frac{[HA]}{C_A} = \frac{[H^+]}{K_a + [H^+]} \tag{5-71}$$

$$\alpha_A = 1 - \alpha_{HA} = \frac{[A^-]}{C_A} = \frac{K_a}{K_a + [H^+]} \tag{5-72}$$

Similarly, for an equilibrium written as an associative interaction, say, $M + L \rightleftharpoons ML$, the fractional distributions of M between the two forms are

$$\alpha_M = \frac{[M]}{C_M} = \frac{1}{1 + K[L]} \tag{5-73}$$

$$\alpha_{ML} = 1 - \alpha_M = \frac{[ML]}{C_M} = \frac{K[L]}{1 + K[L]} \tag{5-74}$$

The usual situation in kinetics is that the variable in *direct control* of the experimenter is the total amount of a substance added, such as C_A or C_M. The pertinent equilibria control the concentrations of the species present, and they must be used to convert the rate law from one containing stoichiometric concentrations to one con-

[†]Stoichiometric concentrations are termed by different workers as laboratory, analytical, formal, or total concentrations.

taining concentrations of species. The conversion is most conveniently accomplished by substitutions of the form shown in Eqs. (5-71) to (5-74).

By way of example, consider the decomposition of perbenzoic acid.

$$2C_6H_5CO_3H = 2C_6H_5CO_2H + O_2 \tag{5-75}$$

The mechanism[7] is proposed to consist of the bimolecular reaction of one molecule of the acid with a second of its conjugate base:

$$C_6H_5CO_3H \rightleftharpoons C_6H_5CO_3^- + H^+ \quad \text{(rapid, p}K_a = 7.78) \tag{5-76}$$

$$C_6H_5CO_3H + C_6H_5CO_3^- \xrightarrow{k_2} \text{products, or intermediate leading}$$
$$\text{rapidly to products} \tag{5-77}$$

If [T] represents the total concentration of perbenzoic acid, the rate law of Eq. (5-78) is easily transformed to one expressed in terms of [T]:

$$\frac{d[O_2]}{dt} = k_2[C_6H_5CO_3H][C_6H_5CO_3^-] \tag{5-78}$$

$$\frac{d[O_2]}{dt} = \frac{k_2 K_a[H^+]}{(K_a + [H^+])^2}[T]^2 = k_{obs}[T]^2 \tag{5-79}$$

Equation (5-79) predicts a bell-shaped variation of log k_{obs} versus pH, maximizing at pH = pK_a, and indeed that is what is observed (Fig. 5-1). The shape of the curve

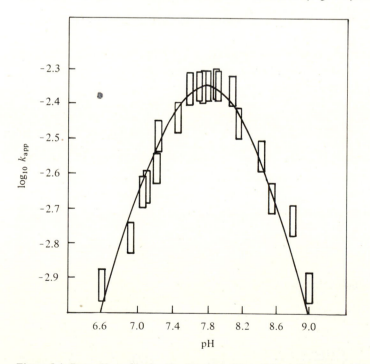

Figure 5-1 Rate-pH profile for the decomposition of peroxobenzoic acid at 25°. Data from Ref. 7.

and the seeming complexity arise from the protonation equilibrium. The actual mechanism is a simple one: bimolecular reaction of acid and base as shown in Eq. (5-77).

A second example will prove instructive, for it illustrates the point that in many circumstances a rate law written in terms of stoichiometric concentrations has the same mathematical form as the rate law applicable to a sequential reaction mechanism.

Taube and Posey[8] found that the rate of the reaction

$$Co(NH_3)_5H_2O^{3+} + SO_4^{2-} = Co(NH_3)_5SO_4^+ + H_2O \tag{5-80}$$

followed pseudo-first-order kinetics in the presence of a large excess of sulfate ions. Above 0.02 M SO_4^{2-}, however, the rate became independent of $[SO_4^{2-}]$ in accord with the equation

$$\frac{d[Co(NH_3)_5SO_4^+]}{dt} = \frac{A[SO_4^{2-}][Co(NH_3)_5H_2O^{3+}]}{1 + B[SO_4^{2-}]} \tag{5-81}$$

Now, if an ion-pairing equilibrium had gone unrecognized, Eq. (5-81) could have been (incorrectly) interpreted in terms of a limiting S_N1 mechanism in which a five-coordinate intermediate $Co(NH_3)_5^{3+}$ plays a major role:

$$Co(NH_3)_5OH_2^{3+} \underset{-1}{\overset{1}{\rightleftharpoons}} Co(NH_3)_5^{3+} + H_2O \tag{5-82}$$

$$Co(NH_3)_5^{3+} + SO_4^{2-} \overset{2}{\longrightarrow} Co(NH_3)_5SO_4^+ \tag{5-83}$$

This formulation is consistent with the experimental equation, with $A = k_1 k_2 / k_{-1}$ and $B = k_2 / k_{-1}$.

In fact, the effect arises from an entirely different phenomenon. Taube and Posey recognized that the reactants associate to an outer-sphere complex or ion pair according to the reaction

$$Co(NH_3)_5H_2O^{3+} + SO_4^{2-} \overset{K_{IP}}{\rightleftharpoons} Co(NH_3)_5H_2O^{3+} \cdot SO_4^{2-} \tag{5-84a}$$

$$[Co(NH_3)_5H_2O^{3+} \cdot SO_4^{2-}] \overset{k_3}{\longrightarrow} Co(NH_3)_5SO_4^+ + H_2O \tag{5-84b}$$

This alters the interpretation, because the experimental rate constant of Eq. (5-81) is (automatically) expressed in terms of total cobalt and must be converted to an expression involving the concentrations of species. Substitution of the relation

$$\frac{[Co(NH_3)_5H_2O^{3+} \cdot SO_4^{2-}]}{[Co]_T} = \frac{K_{IP}[SO_4^{2-}]}{1 + K_{IP}[SO_4^{2-}]} \tag{5-85}$$

yields the equivalent kinetic expressions

$$\frac{d[Co(NH_3)_5SO_4^+]}{dt} = A[Co(NH_3)_5H_2O^{3+}][SO_4^{2-}] \tag{5-86a}$$

$$\frac{d[Co(NH_3)_5SO_4^+]}{dt} = \frac{A}{K_{IP}}[Co(NH_3)_5H_2O^{3+} \cdot SO_4^{2-}] \tag{5-86b}$$

and identifies parameter B of Eq. (5-81) as K_{IP}. This new rate law, which expresses the rate in terms of the concentrations of species present in solution, no longer displays the feature of the limiting rate in SO_4^{2-}.

Thus we see the difference between the situation in which an association or dissociation equilibrium converts a *significant* fraction of the reactant into a different form, one which may be more or less reactive than the original, and the case in which only a minute fraction is converted to *highly reactive* form. These cases, of quite great mechanistic difference, often lead to rate laws of the same algebraic form. In general, one must be able to determine the quantitative extent of the interaction to distinguish between the alternatives. Experiments having higher concentrations of cobalt complex would have served to distinguish the possibilities. In these experiments, if the ion-pairing mechanism holds, $[SO_4^{2-}]$ can no longer be approximated as $[SO_4^{2-}]_T$, whereas the relation remains valid for the S_N1 mechanism if no appreciable quantity of ion pair forms. There may well be practical limitations to the application of this method, however, arising from the more complex analysis required for the kinetic data during each run where an irreversible pseudo-first-order relation will no longer apply.

Still further consideration of this reaction may be instructive. Suppose that the mechanism *were* the S_N1 process, with the dissociative reaction (5-82) rate-limiting and (5-83) quite fast ($k_2[SO_4^{2-}] \gg k_{-1}$). Suppose also that the ion-pairing equilibrium of (5-84a) were important but nonproductive (i.e., the ion pair forms to an extent given by this equilibrium, but it is nonreactive, $k_3 = 0$). In this case it is easily shown, for experiments having a large excess of sulfate ions, that

$$\frac{d[Co(NH_3)_5SO_4^+]}{dt} = \frac{k_1[Co(NH_3)_5H_2O^{3+}]_T[SO_4^{2-}]}{1 + K_{IP}[SO_4^{2-}]}$$

The fact that the coefficient of the sulfate ion term in the denominator would have the numerical value given by an independent determination of K_{IP} can be seen from this equation to be neither coincidental nor indicative of a key role of ion pairing in the mechanism. It merely reflects, in this case, the extent of conversion of the cobalt complex to what is presumed in this mechanism to be a nonreactive storage point.

Consider once again the mechanism of the reaction between V^{3+} and Cr^{2+}, Eqs. (5-52) and (5-53). One might argue for a mechanism far simpler than those proposed in mechanisms I and II. For example, consider the following possibility:

$$V^{3+} + H_2O \rightleftharpoons VOH^{2+} + H^+ \quad \text{(rapid, } K_a\text{)} \tag{5-87}$$

$$VOH^{2+} + Cr^{2+} \xrightarrow{k_s} CrOH^{2+} + V^{2+} \tag{5-88}$$

$$CrOH^{2+} + H^+ = Cr^{3+} + H_2O \tag{5-89}$$

Accounting for a distribution of vanadium(III) between the two forms,

$$[VOH^{2+}] = \frac{[V(III)]_T K_a}{(K_a + [H^+])} \tag{5-90}$$

the rate law associated with this mechanism is

$$\frac{d[V^{2+}]}{dt} = k_5 [VOH^{2+}][Cr^{2+}] \tag{5-91}$$

$$\frac{d[V^{2+}]}{dt} = \frac{k_5 K_a [V(III)]_T [Cr^{2+}]}{[H^+] + K_a} \tag{5-92}$$

This form is *algebraically* equivalent to that observed experimentally, with $q = k_5 K_a$ and $r = K_a$. The basis for discarding the mechanism is that the experimental value of r (0.108 M) and the known value of K_a ($\sim 2 \times 10^{-3}$ M) agree far too poorly to be reconciled. Discarding the mechanism does not, of course, preclude the involvement of VOH^{2+} as the reactive species in other, possibly correct, mechanisms, such as that given earlier in mechanism II (page 100)

REFERENCES

1. Taube, H.: *J. Chem. Educ.*, 36:451 (1959).
2. Hicks, K. W., and J. P. Sutter: *J. Phys. Chem.*, 75:1107 (1971).
3. Haim, A.: *Inorg. Chem.*, 5:2081 (1966).
4. Espenson, J. H.: *Inorg. Chem.*, 4:1025 (1966).
5. Newton, T. W., and F. B. Baker: *Advances in Chem. Ser.*, 71:268 (1967).
6. Birk, J. P.: *J. Chem. Educ.*, 47:805 (1970).
7. Goodman, J. F., P. Robson, and E. R. Wilson: *Trans. Faraday Soc.*, 58:1846 (1962).
8. Taube, H., and F. A. Posey: *J. Am. Chem. Soc.*, 75:1463 (1953).

PROBLEMS

5-1 *Multistep reactions; limiting forms; equivalent mechanisms.* Consider scheme I for the net reaction $2Fe^{2+} + Tl^{3+} = 2Fe^{3+} + Tl^+$:

Scheme I:
$$Tl^{3+} + Fe^{2+} \underset{-1}{\overset{1}{\rightleftharpoons}} Tl^{2+} + Fe^{3+}$$

$$Tl^{2+} + Fe^{2+} \overset{2}{\longrightarrow} Tl^+ + Fe^{3+}$$

(a) Derive the rate expression making the steady-state approximation for $[Tl^{2+}]$.

(b) To what simpler forms does this reduce under what limiting conditions?

(c) Are, and under what conditions, schemes II and III distinguishable from I?

Scheme II:
$$Tl^{3+} + Fe^{2+} \underset{-3}{\overset{3}{\rightleftharpoons}} Tl^+ + Fe^{4+}$$

$$Fe^{4+} + Fe^{2+} \overset{4}{\longrightarrow} 2Fe^{3+}$$

Scheme III:
$$2Fe^{2+} \rightleftharpoons Fe^+ + Fe^{3+} \quad (K_5 \ll 1)$$

$$Fe^+ + Tl^{3+} \underset{-6}{\overset{6}{\rightleftharpoons}} Fe^{2+} + Tl^{2+}$$

$$Fe^{3+} + Tl^{2+} \overset{7}{\longrightarrow} Fe^{4+} + Tl^+$$

$$Fe^{4+} + Fe^{2+} = 2Fe^{3+} \quad \text{(very fast)}$$

5-2 *Rate law and mechanism.* Cetini et al. [*Inorg. Chem.*, **10**:2672 (1971)] propose this mechanism:

$$[Ni(cp)_2(CO)_2] \underset{-1}{\overset{1}{\rightleftarrows}} [Ni(cp)_2(CO)_2]^*$$

$$[Ni(cp)_2(CO)_2]^* + PhCCPh \xrightarrow{2} [Ni(cp)_2(PhCCPh)] + 2CO$$

Net: $[Ni(cp)_2(CO)_2] + PhCCPh = [Ni(cp)_2(PhCCPh)] + 2CO$

in which $[Ni(cp)_2(CO)_2]^*$ represents a steady-state intermediate of rearranged structure.

(*a*) Derive the differential rate expression. Designate the pseudo-first-order rate constant k_{obs}. Prove that $1/k_{obs}$ versus $1/[PhCCPh]$ should be linear (i.e., $1/k_{obs}$ = intercept + slope/[PhCCPh]).

(*b*) Assuming each of the three individual rate constants follows the Arrhenius equation, $k_i = A_i \exp(-E_i/RT)$, identify the exact quantity given by the slope of the following plots:

1. ln (1/intercept) versus $1/T$
2. ln (slope/intercept) versus $1/T$

5-3 *Rate law and mechanism.* Write a mechanism consistent with the following observations, and give the algebraic relation between the constants of your mechanism and those in the experimental rate law. Mawby and coworkers [*J. Chem. Soc. Dalton*, **220** (1973)] report that the isomerization of A to B

follows the rate expression (L is a phosphine, PMe_2Ph).

$$\frac{d[B]}{dt} = \frac{P[A]}{1 + Q[L]}$$

5-4 *Rate law and mechanism.* The redistribution of alkyl groups on silanes as in the equation

$$2Me_3SiEt = Me_2SiEt_2 + Me_4Si$$

is catalyzed in benzene solution by aluminum bromide according to the rate expression

$$\frac{-d[Me_3SiEt]}{dt} = \frac{\alpha[Me_3SiEt]^2[Al_2Br_6]}{1 + \beta[Me_3SiEt]}$$

Propose a mechanism to account for this result and show how the rate constants for the elementary reactions are related to α and β.

5-5 *Rate law and mechanism.* The net reaction

$$2V(III) + 2Hg(II) = 2V(IV) + [Hg(I)]_2$$

has a rate term showing the following concentration dependences

$$\frac{-d[V(III)]}{dt} = \frac{A[V(III)]^2[Hg(II)]}{B[V(IV)] + [V(III)]}$$

Suggest a mechanism consistent with this rate expression noting that on the time scale of this reaction the following are fast and lie far to the right

$$V(III) + V(V) \longrightarrow 2V(IV) \qquad Hg(II) + Hg^0 \longrightarrow [Hg(I)]_2$$

5-6 *Rate law and mechanism.* P. C. Ellgen [*Inorg. Chem.*, **11**:691 (1972)] has studied the reaction of dicobalt octacarbonyl with alkynes:

$$Co_2(CO)_8 + RC_2R' = Co_2(CO)_6RC_2R' + 2CO$$

under conditions in which $[RC_2R']$ and $[CO]$ were in large excess over $[Co_2(CO)_8]$. The symbol k_{obs} represents the pseudo-first-order rate constant defined by the relation $-d \ln [Co_2(CO)_8]/dt$. Consider the following two mechanisms:

Mechanism I:
$$Co_2(CO)_8 \underset{k_{-1}}{\overset{k_1}{\rightleftharpoons}} Co_2(CO)_7 + CO$$

$$Co_2(CO)_7 + Ph_2C_2 \xrightarrow{k_2} Co_2(CO)_6Ph_2C_2 + CO$$

Mechanism II:
$$Co_2(CO)_8 + Ph_2C_2 \underset{k_{-3}}{\overset{k_3}{\rightleftharpoons}} Co_2(CO)_7Ph_2C_2 + CO$$

$$Co_2(CO)_7Ph_2C_2 \xrightarrow{k_4} Co_2(CO)_6Ph_2C_2 + CO$$

(a) For each mechanism derive an expression for k_{obs}. Make the steady-state assumption for the intermediate.

(b) What (be specific) do the following observations reveal about the correctness of either mechanism? k_{obs} is a linear function of $1/[C_2Ph_2]_0$ at constant $[CO]$, and the slope of the plot varies directly with $[CO]$. Provide numerical values for any rate constants or rate constant combination you can on the basis of these data.

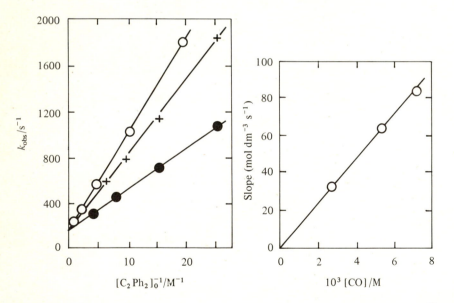

5-7 *Reaction mechanism.* The reaction of dichromate ion with dihydrogen, $Cr_2O_7^{2-} + 3H_2 + 8H^+ = 2Cr^{3+} + 7H_2O$, is very slow, but it is catalyzed by salts of Cu^{2+} and Ag^+ according to the following rate laws [J. Halpern et al., *J. Phys. Chem.*, 60:1455 (1956); 61:1239 (1957)]:

Cu²⁺ catalysis:
$$\frac{-d[Cr_2O_7^{2-}]}{dt} = \frac{k_{Cu}[Cu^{2+}]^2[H_2]}{[H^+]}$$

Ag⁺ catalysis:
$$\frac{-d[Cr_2O_7^{2-}]}{dt} = k_{1Ag}[Ag^+]^2[H_2] + k_{2Ag}[Ag^+][H_2]$$

Interpret (separately) these observations in terms of reaction mechanisms.

5-8 *Rate law and mechanism.* Propose a mechanism for each and relate the parameters of the rate law to the rate constants for the mechanism.

(*a*) The hydrolysis of benzhydryl chloride in aqueous acetone,

$$Ph_2CHCl + H_2O = Ph_2CHOH + Cl^- + H^+; \quad \frac{d[Ph_2CHOH]}{dt} = \frac{\alpha[Ph_2CHCl]}{\beta + [Cl^-]}$$

(*b*) The oxidation of iodide by aqueous Fe^{3+}, $2Fe^{3+} + 3I^- = 2Fe^{2+} + I_3^-$,

$$\frac{-d[Fe^{3+}]}{dt} = \frac{k[Fe^{3+}]^2[I^-]^2}{[Fe^{3+}] + k'[Fe^{2+}]}$$

(*c*) Smith and Yates [*J. Am. Chem. Soc.*, **94**:8811 (1972)] find that the hydrolysis of tri-methylbenzimidate (I) follows the rate law

$$\frac{-d[I]}{dt} = \frac{A + B/[H^+]}{1 + C[H^+]}[I]$$

At low $[H^+]$ the products are primarily amine plus ester, whereas at high $[H^+]$, amide and alcohol are formed:

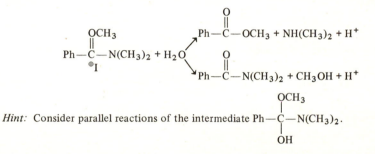

Hint: Consider parallel reactions of the intermediate $Ph-\underset{\underset{OH}{|}}{\overset{\overset{OCH_3}{|}}{C}}-N(CH_3)_2$.

5-9 *Mechanistic distinction and interpretations; preequilibria.* The oxidation of I^- by $HCrO_4^-$ is *catalyzed by* oxalic acid. Vandegrift and Rocek [*J. Am. Chem. Soc.*, **98**:1372 (1976)] have carried out studies under conditions in which $[HCrO_4^-] \ll [H^+]$, $[I^-]$, and $[(COOH)_2]$. The rate is first-order in Cr(VI), $-d[Cr(VI)]/dt = k_{cat}[Cr(VI)]$.

The authors formulate two alternative mechanisms: I, in which a rapid and reversible equilibrium converts a significant portion of $HCrO_4^-$ to $ICrO_3^-$, and II, in which Cr(VI) exists only as $HCrO_4^-$ and reacts in a two-step sequence involving a steady-state oxalochromate intermediate. Derive the rate law for each mechanism. A plot of $1/k_{cat}$ versus $1/[I^-]$ (at constant $[(COOH)_2]$ and $[H^+]$) is linear, with intercept designated $1/A$ and intercept/slope = B. The data are shown below; interpret in terms of schemes I and/or II.

Scheme I:
$$HCrO_4^- + I^- + H^+ \overset{K_1}{\rightleftharpoons} ICrO_3^- + H_2O$$

$$ICrO_3^- + (CO_2H)_2 \xrightarrow{2} \text{products}$$

Scheme II:
$$HCrO_4^- + (CO_2H)_2 + H^+ \underset{-3}{\overset{3}{\rightleftharpoons}} \quad + 2H_2O$$

$$(CO_2)_2CrO_2 + I^- \xrightarrow{4} \text{products}$$

$[H^+]$	$10^3 [(CO_2H)_2]$	A	B
0.030	0.50	5.88	1.12
0.030	10.00	118	1.24
0.100	2.00	25.0	4.0
0.100	3.00	38.8	3.75
0.100	15.00	192	4.30
0.300	2.00	25.3	11.7
0.300	5.00	58.8	11.3
0.600	5.00	66.7	21.8

5-10 *Preequilibria and reaction mechanism.* Woodruff, Weatherburn, and Margerum [*Inorg. Chem.*, **10**:2102 (1971)] have studied the oxidation of certain Fe(II) complexes (Fe^{II}L) by iodine-triiodide solutions.

$$2Fe^{II}L + I_3^- = 2Fe^{III}L + 3I^-$$

The rate is first-order in [Fe^{II}L] and first-order in total iodine concentration $[I_2]_T$ (where $[I_2]_T = [I_2] + [I_3^-]$):

$$\frac{-d[I_2]_T}{dt} = k_0[Fe^{II}L][I_2]_T$$

A correction was applied for the equilibrium

$$I_2 + I^- \rightleftharpoons I_3^- \qquad (K_{I_3} = 770 \text{ M}^{-1} \text{ at } 25°)$$

Prove that the data shown are consistent with parallel rate-limiting reactions of I_2 and I_3^- and evaluate k_1 and k_2 at 25.0°.

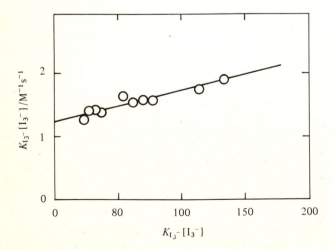

5-11 *Preequilibria.* The reactions of cyanopyridine complexes of cobalt(III) with chromium(II) have been studied [*J. Am. Chem. Soc.*, **98**:1487 (1976)]. The reactions are complicated by a protonation

$$\left[(NH_3)_5Co-N\equiv C-\underset{\underset{H}{N}}{\bigcirc} \right]^{4+} \xrightarrow{K_a} \left[(NH_3)_5Co-N\equiv C-\underset{N}{\bigcirc} \right]^{3+} + H^+$$

(B)

Derive the rate expression for $-d[Co(III)$ complex$]/dt$ in these cases: (*a*) The 2-cyanopyridine complex shown, in which only the basic form reacts with Cr^{2+}, $B_2 + Cr^{2+} \xrightarrow{k_1}$ products, and (*b*) The analogous 3-cyanopyridine complex in which parallel reaction of acid and base occur: $A_3 + Cr^{2+} \xrightarrow{k_2}$ products and $B_3 + Cr^{2+} \xrightarrow{k_3}$ products.

5-12 *Preequilibria.* Biechler and Taft [*J. Am. Chem. Soc.*, **79**:4932 (1957)] have studied the base hydrolysis of trifluoroacetanalide, a reaction which is complicated by an acid-base equilibrium:

Net reaction:
$$CF_3\overset{O}{\overset{\|}{C}}NPhH + OH^- = CF_3CO_2^- + PhNH_2$$

Acid base:
$$CF_3CONPhH \rightleftharpoons CF_3CONPh^- + H^+ \qquad pK_a = 12.43$$

The reaction is pseudo-first-order in the total concentration of reactant, and shows the pH profile depicted. Give a complete mechanistic interpretation.

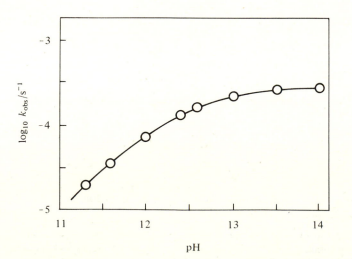

5-13 *Preequilibria.* Hutchins and Fife [*J. Am. Chem. Soc.*, 95:2282 (1973)] have shown that carbamate XOH undergoes an instantaneous acid-base equilibrium

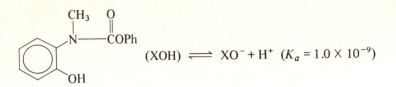

$$(XOH) \rightleftharpoons XO^- + H^+ \quad (K_a = 1.0 \times 10^{-9})$$

but is susceptible to decomposition:

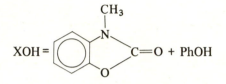

$$XOH = \qquad C=O + PhOH$$

Give a quantitative interpretation of the kinetic data. The rate law is given by

$$\frac{d[PhOH]}{dt} = k_{obs}[X]_T$$

where $[X]_T = [XOH] + [XO^-]$, and the pH profile is as shown.

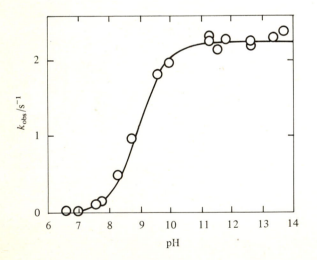

5-14 *Composite rate constants.* Give the expression by which k_4 of Eq. (5-20) can be calculated from the constants of Eqs. (5-15) to (5-19).

5-15 *Preequilibria and product distribution.* Benkovic et al. [*J. Am. Chem. Soc.*, 95:8414 (1973)] have studied the hydrolysis of formamide III*a* to a mixture of N-1 and N-10 formyl products; the reactant is involved in an ionization equilibrium ($K_a = 2 \times 10^{-9}$ M):

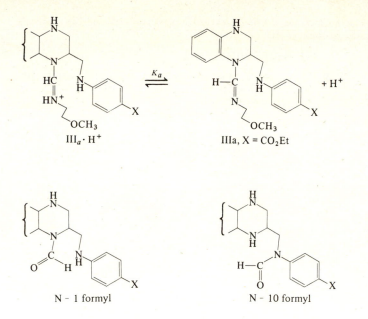

III$_a$·H$^+$ IIIa, X = CO$_2$Et

N - 1 formyl   N - 10 formyl

They evaluate the pseudo-first-order rate constant and the percentage yield of N-1 product both as functions of pH, obtaining data depicted below. Propose a mechanism; show that it accounts for these results; and evaluate all possible rate constants.

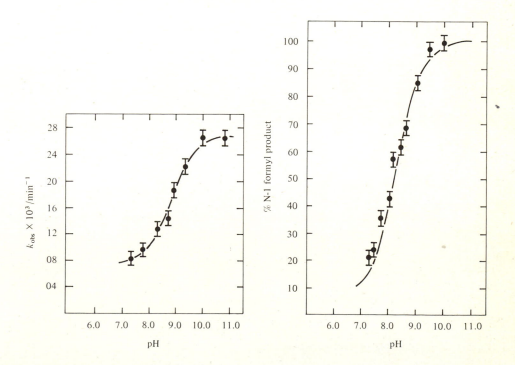

5-16 *Preequilibria and rate law.* The rate of isotopic exchange between U(IV) and U(VI) follows a second-order rate expression at constant $[H^+]$:

$$R_{ex} = k_{ex}[U(VI)][U(IV)]_T$$

where $[U(IV)]_T = [U^{4+}] + [UOH^{3+}]$, related by the equilibrium $U^{4+} + H_2O = UOH^{3+} + H^+$, $K_a = 5.6 \times 10^{-2}$ mol dm^{-3}. The only U(VI) species in the pH range considered is UO_2^{2+}. Using the data provided [from Schwartz and Masters, *J. Am. Chem. Soc.*, 83:2620 (1961)], express the exchange rate as a function of $[UO_2^{2+}]$, $[U^{4+}]$, and $[H^+]$, and compute the rate constant(s) of this equation.

$[H^+]$/mol dm^{-3}	0.200	0.100	0.0500	0.0250
k_{ex}/dm^3 mol^{-1} s^{-1}	2.05×10^{-5}	1.34×10^{-4}	7.92×10^{-4}	4.15×10^{-3}

5-17 *pH profile and reaction mechanism.* Switkes, Dasch, and Ackerman [*Inorg. Chem.*, **12**:1120 (1973)] have studied the decomposition of *N*-nitrosohydroxylamine-*N*-sulfonate ion (= X^{2-}).

$$\left[O=N-N \begin{array}{c} SO_3^- \\ \diagdown \\ O^- \end{array} \right] = SO_4^{2-} + N_2O$$

The rate law follows the expression

$$\frac{-d[X^{2-}]}{dt} = k_{obs}[X^{2-}]$$

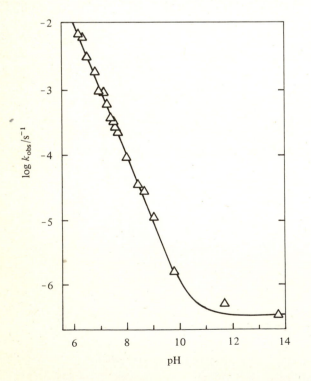

The pH profile at 25°C is shown. The equilibrium $HX^- = X^{2-} + H^+$ has $K_a = 1.0 \times 10^{-2}$ mol dm^{-3}. Formulate a mechanism consistent with these data.

5-18 *Reaction mechanism.* Interpret the following data as completely as possible. S. J. Okrasinski and J. R. Norton [*J. Am. Chem. Soc.*, 99:295 (1977)] have studied two reactions of $HOs(CO)_4CH_3$, thermolysis(1) and reaction with phosphine(2):

1: $$2HOs(CO)_4CH_3 = CH_4 + HOs(CO)_4Os(CO)_4CH_3$$

2: $$HOs(CO)_4CH_3 + PR_3 = CH_4 + Os(CO)_4PR_3$$

The reaction rates at 49°C in methylcyclohexane are:

Thermolysis: $$\frac{-d[HOs(CO)_4CH_3]}{dt} = k_1[HOs(CO)_4CH_3]$$

$$k = (1.38 \pm 0.06) \times 10^{-4} \text{ s}^{-1}$$

Phosphine: $$\frac{-d[HOs(CO)_4CH_3]}{dt} = k_2[HOs(CO)_4CH_3]$$

$$k_2 = (6.4 \pm 0.1) \times 10^{-5} \text{ s}^{-1}$$

A mixture of $HOs(CO)_4CD_3$ and $DOs(CO)_4CH_3$, in either thermolysis or reaction with phosphine, yields a mixture of CD_3H and CH_3D, but no CD_4 or CH_4.

5-19 *pH depenedence.* Williams and Douglas [*J. Chem. Soc. Perkin II*, 1727 (1974)] have studied the kinetics of the following hydrolysis reaction:

$$NO_2-\langle\bigcirc\rangle-OSO_2NHMe + H_2O = NO_2-\langle\bigcirc\rangle-OH + MeNHSO_3^- + H^+$$

The starting material undergoes acid dissociation with pK_a 8.9; the reaction proceeds by parallel hydrolysis of acid ($k_1 = 4 \times 10^{-6}$ s^{-1}) and anion ($k_2 = 8 \times 10^{-3}$ s^{-1}). Sketch the expected pH profile showing the important features semiquantitatively with numerical scales on the axes.

SIX

REACTION ENERGETICS AND CHEMICAL KINETICS

There are a number of fronts on which considerations of kinetics and thermodynamics merge, most notably in the analysis of a rate constant in terms of the energetics of the activation process. The aspects of the interrelations considered in the present chapter are these: (1) the temperature dependence of the rate constant, (2) the relations between forward and reverse reaction rates, and (3) the principle of microscopic reversibility, especially its implications for reaction mechanisms.

6-1 THE VARIATION OF RATE CONSTANT WITH TEMPERATURE

The temperature dependence of the rate constant for an elementary reaction is almost always expressed by a relation of the form given by Eq. (6-1). The quantity U has the

$$k = CT^n \exp\left(-\frac{U}{RT}\right) \tag{6-1}$$

dimensions of energy, and the exponent n is usually assigned a value of 0, $\frac{1}{2}$, or 1 depending upon the theoretical model chosen. Once n is chosen, a plot of the data as $\ln kT^{-n}$ against T^{-1} should be linear with a slope of $-U/R$ as seen by the relation

$$\ln \frac{k}{T^n} = \ln C - \frac{U}{R} T^{-1} \tag{6-2}$$

The value of C is then easily calculated by substitution. Or perhaps more simply, the data may be fit to Eq. (6-1) directly by the method of least squares. By such methods one obtains the values of the two adjustable parameters C and U. The three variants of Eq. (6-1) most commonly encountered are these:

(a) The Arrhenius relation, Eq. (6-3), in which the two parameters are A, the

$$k = A \exp\left(-\frac{E_a}{RT}\right) \tag{6-3}$$

preexponential factor (or frequency factor, the latter being a poor name because only in the case of a unimolecular reaction does A have frequency dimensions of s^{-1}), and E_a, termed the Arrhenius activation energy or sometimes simply the activation energy. The preexponential factor A is taken as independent of temperature.

(b) The collision theory relation for bimolecular reactions in the gas phase is given by Eq. (6-4), in which p symbolizes the so-called steric factor, E^* is the critical

$$k = pZ \exp\left(-\frac{E^*}{RT}\right) \tag{6-4}$$

collision energy, and Z is the collision frequency at unit concentration of reactants. The value of Z is obtained from the kinetic molecular theory of gases; it depends on the collision distance d_{AB} and on the masses of the reactants A and B (Sec. 8-1) according to Eq. (6-5). The preexponential value is thus proportional to $T^{1/2}$.

$$Z_{AB} = \left[\frac{8\pi kT}{m_A m_B}(m_A + m_B)\right]^{1/2} d_{AB}{}^2 \tag{6-5}$$

(c) The Eyring or activated complex or absolute rate theory relation employs, as do the other formulations, two adjustable parameters. In this case they are $\Delta H^{\ddagger}$ the activation enthalpy and $\Delta S^{\ddagger}$ the activation entropy. The preexponential shows a first-power temperature dependence, Eq. (6-6), in which transmission coefficient κ is usually taken as unity

$$k = \kappa \frac{RT}{Nh} \exp\left(\frac{\Delta S^{\ddagger}}{R}\right) \exp\left(-\frac{\Delta H^{\ddagger}}{RT}\right) \tag{6-6}$$

The theories will be considered in their own right in Secs. 8-2 and 8-3. For the present, the mathematical statements in Eqs. (6-5) and (6-6) will simply be taken as relations which can be used to correlate experimental values of a rate constant k as a function of temperature.

It may prove difficult to see at the outset why it is that the appropriate power of T in the preexponential term cannot be decided by accurate measurements. The difficulty is that the exponential energy term is usually so dominant that the differences in taking $n = 0$, $\frac{1}{2}$, or 1 are negligible. That is to say, for any reaction in which a plot of $\ln k$ versus T^{-1} is linear as required by the Arrhenius relation, plots of $\ln (k/T^{1/2})$ and of $\ln (k/T)$ versus T^{-1} are also linear within the usual experimental error of the kinetic data. To illustrate, consider the supposedly typical reaction whose rate at 300 K roughly doubles over a $10°$ temperature increase. Such a reaction has $U \sim 50$ kJ mol^{-1}.[†] The ratio of rate constants for the 10° increase is

$$\frac{k_{310}}{k_{300}} = \left(\frac{310}{300}\right)^n \exp\left[\frac{U}{R}\left(\frac{1}{300} - \frac{1}{310}\right)\right] \tag{6-7}$$

With the values specified above, k_{310}/k_{300} is 1.91 ($n = 0$), 1.94 ($n = \frac{1}{2}$), or 1.97 ($n = 1$). Clearly the exponential term is dominant, and once the choice of n has been made, only a slightly different value of U is required to correlate the data.

To dwell on this point but a bit longer, suppose one set out to decide the "correct" preexponential expression by using rate data of exceptionally high precision in a reaction of very small activation energy and thereby seeking to minimize the importance of the energy exponential. It is unlikely that would provide a definitive answer, because the equations as they are normally used take U as being temperature-independent. That is akin to taking a thermodynamic quantity $\Delta H°$ as temperature-independent, or $\Delta Cp° = 0$, a good approximation but insufficient for high accuracy. The assumption of the temperature invariance of U introduces an assumption of the same magnitude as that contained in the question of the temperature exponent n. On the other hand, to explicitly include a temperature dependence, say a nonzero value of $\Delta C_p^{\ddagger}$, is to render the value of n indeterminate, owing to the one or more additional adjustable parameters contained in $\Delta C_p^{\ddagger}$.

6-2 ACTIVATION PARAMETERS

As an example of the calculations involved, consider data for the bimolecular reaction[1] of Eq. (6-8), for which rate constants at various temperatures are

$$NO + ClNO_2 \longrightarrow NO_2 + ClNO \tag{6-8}$$

T/K	300	311	323	334	344
$10^{-7} \ k/cm^3 \ mol^{-1} \ s^{-1}$	0.79	1.25	1.64	2.56	3.4

Values of $\ln k$, $\ln (k/T^{1/2})$, or $\ln (k/T)$ are plotted versus $1/T$; the slope, $-U/R$, is computed from a graphical or numerical analysis. These graphs are shown in Fig. 6-1. The numerical parameters from this treatment are incorporated into the following expressions:

Arrhenius:

$$k = 6.5 \times 10^{11} \exp\left(-\frac{28.2 \ kJ \ mol^{-1}}{RT}\right) \tag{6-9}$$

[†]A note on SI units: Convert (if you must) from SI units as follows: kilojoules to kilocalories by division by 4.184; 50 kJ mol^{-1} $\approx$ 12 kcal mol^{-1}. The constant $R = 1.987$ cal mol^{-1} K^{-1} = 8.314 J mol^{-1} K^{-1}. Absolute temperatures are denoted as 300 K ("300 kelvins").

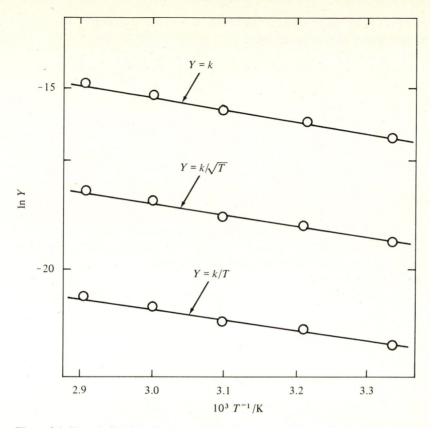

Figure 6-1 Plots indicating the temperature dependence of the rate constants. Upper line, Arrhenius relation, ln k versus $1/T$; middle line, collision theory, ln $(k/T^{1/2})$ versus $1/T$; lower line, activated complex theory, ln (k/T) versus $1/T$.

Collision theory:

$$k = 2.1 \times 10^{10}\ T^{1/2}\ \exp\left(-\frac{26.8\ \text{kJ mol}^{-1}}{RT}\right) \tag{6-10}$$

Activated complex theory:

$$k = 7.1 \times 10^{8}\ T\ \exp\left(-\frac{25.4\ \text{kJ mol}^{-1}}{RT}\right) \tag{6-11}$$

The numerical value 7.1×10^{8} must be equal to the quantity $(R/Nh)\ \exp\ (\Delta S^{\ddagger}/R)$ of Eq. (6-6). With substitution of the values for the physical constants, the numerical value of $\Delta S^{\ddagger}$ is -28.1 J mol^{-1} K^{-1}, and the rate constant expressed in the form of Eq. (6-6) is

$$k = \frac{RT}{Nh}\ \exp\left(-\frac{28.1\ \text{J mol}^{-1}\ \text{K}^{-1}}{R}\right) \times \exp\left(-\frac{25.4\ \text{kJ mol}^{-1}}{RT}\right) \tag{6-12}$$

If the rate constant values are first converted to the standard state[†] 1 mol dm^{-3}, the absolute rate theory expression becomes

$$k = \frac{RT}{Nh} \exp\left(-\frac{85.5 \text{ J mol}^{-1} \text{ K}^{-1}}{R}\right) \times \exp\left(-\frac{25.4 \text{ kJ mol}^{-1}}{RT}\right) \qquad (6\text{-}13)$$

The value of $\Delta S^{\ddagger}$ changes from -28.1 to -85.5 J mol^{-1} K^{-1}. The magnitude of the change is that due solely to the "dilution" implied by the change in standard state.[§]

It is this author's view that nothing is to be gained by mixing the three expressions. A hybrid of the Arrhenius and activated complex theory expressions, for example, can be obtained by algebraic substitutions, but there is no apparent advantage in using such an equation to correlate kinetic data in terms of, say, E_a and $\Delta S^{\ddagger}$.

All of the expressions considered in this section are properly applied only to an authentic rate constant. Values of apparent rate constants which contain unaccounted for concentration dependences cannot correctly be used. The preexponential factor, in particular, would be erroneous if, say, a pseudo-first-order rate constant were correlated by any of the forms considered.

It should be evident that the numerical values of the activation parameters based on these different equations are related to each other. Consider one case, the relation between E_a of the Arrhenius equation and $\Delta H^{\ddagger}$ of activated complex theory. The defining relations from Eqs. (6-3) and (6-5) are

$$E_a = -R \frac{d \ln k}{d(1/T)}$$

$$\Delta H^{\ddagger} = -R \frac{d \ln (k/T)}{d(1/T)}$$

The latter can be rewritten as

$$\Delta H^{\ddagger} = -R \frac{d \ln k}{d(1/T)} - R \frac{d \ln (1/T)}{d(1/T)}$$

which transforms to

$$\Delta H^{\ddagger} = E_a - RT \qquad (6\text{-}14)$$

Thus, near room temperature, E_a is roughly 2.5 kJ mol^{-1} larger than $\Delta H^{\ddagger}$. By similar methods the relation between A and $\Delta S^{\ddagger}$ can be obtained (see Prob. 6-8).

[†]The concentration unit for species in solution is almost invariably taken as 1 mol dm^{-3}. The situation for gas-phase reactions seems less uniform. The choice of the concentration unit is equivalent to the choice of a standard state for the activation parameters. It is important to note that the numerical value of $\Delta S^{\ddagger}$ (or of any other preexponential factor) varies with this choice, except for first-order reactions.

[§]The bimolecular rate constant corresponds to this process,

$$A + B \longrightarrow [\text{activated complex}]^{\ddagger}$$

The change in $\Delta S^{\ddagger}$ is identical with the changes in ΔS° for a gas-phase equilibrium of the same form, $A + B \rightleftharpoons C$. One mole of an ideal gas expanded from 1 cm^3 to 1 dm^3 increases its entropy by $R \ln (10^3/1) = 57.4$ J mol^{-1} K^{-1}.

6-3 THE TEMPERATURE DEPENDENCE OF COMPOSITE RATE CONSTANTS

It is common to encounter rate expressions in which the experimental rate constant is a composite of the rate constants and/or equilibrium constants for several reactions. Depending on the particular mathematical form of a given expression, such an experimental rate constant may or may not depend on temperature in such a manner that $\ln (k/T^n)$ versus $1/T$ will be linear.

By way of example, consider three mechanisms for the net reaction $A = C$.

(*a*)

$$A \rightleftharpoons B \quad \text{fast}, K_1 \ll 1 \tag{6-15}$$

$$B \xrightarrow{\;2\;} C \quad \text{rate-limiting} \tag{6-16}$$

$$\frac{-d[A]}{dt} = k_2 K_1 [A] \tag{6-17}$$

The form of the temperature dependence is seen from the following formulation

$$k_{app} = k_2 K_1 = \left(\frac{RT}{Nh}\right) \exp\left(\frac{\Delta S_2^{\ddagger}}{R}\right) \exp\left(-\frac{\Delta H_2^{\ddagger}}{RT}\right) \times \exp\left(\frac{\Delta S_1^{\circ}}{R}\right) \exp\left(-\frac{\Delta H_1^{\circ}}{RT}\right)$$

Combining terms gives

$$\frac{k_{app}}{T} = \frac{k_2 K_1}{T} = \frac{RT}{Nh} \exp\left(\frac{\Delta S_1^{\circ} + \Delta S_2^{\ddagger}}{R}\right) \times \exp\left(-\frac{\Delta H_1^{\circ} + \Delta H_2^{\ddagger}}{RT}\right) \tag{6-18}$$

Clearly, a plot of $\ln (k_{app}/T)$ versus $1/T$ will be linear, and the apparent activation enthalpy, computed as $-R \times \text{slope}$, is seen to be the sum of ΔH_1° and $\Delta H_2^{\ddagger}$. It should be noted that if ΔH_1° for the equilibrium step is negative, the observed activation enthalpy will be lower than the value of $\Delta H^{\ddagger}$ for the rate-limiting step. Indeed, occasionally a negative value of $\Delta H_{obs}^{\ddagger}$ may result, and the reaction rate actually decreases with increasing temperature.

(*b*) This mechanism consists of parallel pathways for the reaction of A:

$$A \xrightarrow{\;3\;} C \quad \text{and} \quad A \xrightarrow{\;4\;} B$$
$$B = C \text{ (fast)}$$

$$\frac{-d[A]}{dt} = (k_3 + k_4)[A] \tag{6-19}$$

If we substitute for k_3 and k_4 of Eq. (6-19) an expression from activated complex theory [Eq. (6-6)], the resulting expression for the experimental composite $k_{(6\text{-}19)} = k_3 + k_4$ has the form

$$\frac{k_{(6\text{-}19)}}{T} = C_1 \exp\left(-\frac{\Delta H_3^{\ddagger}}{RT}\right) + C_2 \exp\left(-\frac{\Delta H_4^{\ddagger}}{RT}\right) \tag{6-20}$$

Based on this relation, try to construct the shape of the usual temperature plot, $\ln (k_{(6\text{-}19)}/T)$ versus $1/T$. A pictorial way of doing so is to appreciate that the rate constant having the larger value of $\Delta H^{\ddagger}$ will rise more steeply with increasing temperature until, at some value of T, it will effectively dominate; by analogy, the rate constant having the lower value of $\Delta H^{\ddagger}$ will dominate at lower temperature. Since the slope of the plot under discussion is $-\Delta H^{\ddagger}/R$, such a plot will always curve *gradually* upward when a composite experimental rate constant is a *sum* of individual constants.

(*c*) This mechanism consists of successive steps:

$$A \underset{-5}{\overset{5}{\rightleftharpoons}} B \overset{6}{\longrightarrow} C \qquad B = \text{steady-state intermediate} \qquad (6\text{-}21)$$

$$\frac{-d[A]}{dt} = \frac{k_5 k_6}{k_{-5} + k_6} [A] \qquad (6\text{-}22)$$

This situation also corresponds to a nonlinear plot of $\ln (k_{(6\text{-}21)}/T)$ versus $1/T$ in the range where k_6 and k_{-5} are comparable. In this case, however, the plot will be curved downward (see Prob. 6-10).

In practice, curvature in plots of $\ln (k_{app}/T)$ versus $1/T$ may arise from causes other than a complex mechanism. That is particularly so for reactions in which the presumed temperature independence of E_a or $\Delta H^{\ddagger}$ (that is, $\Delta Cp^{\ddagger} = 0$) is not a good approximation. It may not be one in solution reactions, particularly with ionic reagents, as considered further in Sec. 6-6.

The enthalpy of activation $\Delta H^{\ddagger}$ is a measure of the height of the energy barrier, particularly bond strengths within and between the reactants, which must be overcome to attain the transition state. The larger the value of $\Delta H^{\ddagger}$, the slower the reaction, other factors being equal. An increase of 5.7 kJ mol^{-1} in $\Delta H^{\ddagger}$ corresponds to a factor of 10 in the rate constant at room temperature. If a reaction occurs with a value of $\Delta H^{\ddagger}$ less than the dissociation energy of a particular group in the molecule, one can conclude that the bond to this group has not broken during the activation process.[†]

The entropy of activation relates to the probability of reaction. It includes contributions from requirements imposed by the orientation and steric bulk of the reactants and by their solvation. A unimolecular reaction may have $\Delta S^{\ddagger}$ near zero owing to the lack of orientational factors. If, however, the activation process involves dissociation, such as bond breaking within the single species, a large and positive value of $\Delta S^{\ddagger}$ is likely. As stated earlier, the numerical value of $\Delta S^{\ddagger}$ for a bimolecular reaction is dependent upon the choice of concentration scale. For rate constants in units dm^3 mol^{-1} s^{-1}, the value of $\Delta S^{\ddagger}$ is likely to be substantially negative, at least if the second-order kinetics is a manifestation of a bimolecular mechanism, not a more complicated scheme. The contributions of $\Delta S^{\ddagger}$ include the negative value resulting from bringing together two separate reactants each with 1 mol dm^{-3} concentration (roughly some -33 J mol^{-1} K^{-1}), as well as further contributions, negative as well, reflecting ionic charges (if any) and particular orientational factors, including loss of internal degrees of freedom in the transition state. The more negative the value of $\Delta S^{\ddagger}$, the lower

---

[†]Although generally true, some caution is called for on two counts: (1) It may be possible to devise a mechanism in which a process having a negative enthalpy partly compensates. (2) For reasons considered in Chap. 7, certain chain reactions constitute exceptions.

the reaction rate, decreasing by a factor of 10 for a decrease of -19 J mol^{-1} K^{-1} in $\Delta S^{\ddagger}$.

6-4 RELATION OF FORWARD AND REVERSE REACTION RATES

The treatment of reversible reactions in Chap. 3 employed, for an elementary reaction, the following relation:

$$K = \frac{k_f}{k_r} \tag{6-23}$$

The validity of (6-23) is apparent, because at equilibrium the net rate of reaction becomes zero.

A slightly more complicated situation applies to the two-step process

$$A \underset{-1}{\overset{1}{\rightleftharpoons}} B \tag{6-24}$$

$$B \underset{-2}{\overset{2}{\rightleftharpoons}} C \tag{6-25}$$

$$\frac{-d[A]}{dt} = k_1[A] - k_{-1}[B] \tag{6-26}$$

$$\frac{d[B]}{dt} = k_1[A] - k_{-1}[B] - k_2[B] + k_{-2}[C] \tag{6-27}$$

At equilibrium each of the rates is zero, and that gives the relation

$$K = \frac{[C]_{eq}}{[A]_{eq}} = \frac{k_1 k_2}{k_{-1} k_{-2}} \tag{6-28}$$

Consider a reaction occurring by parallel pathways. The formation and dissociation of azidoiron(III) ion serves as an example[2] [Eq. (6-29)]. In terms of the

$$Fe^{3+} + HN_3 = FeN_3^{2+} + H^+ \tag{6-29}$$

predominant species in solution the net reaction rate is given by

$$\frac{d[FeN_3^{2+}]}{dt} = \left(k_1 + \frac{k_2}{[H^+]}\right)[Fe^{3+}][HN_3] - (k_{-1}[H^+] + k_{-2})[FeN_3^{2+}] \tag{6-30}$$

By equating forward and reverse rates at equilibrium and rearranging the resulting concentrations to correspond with the correct form of the equilibrium constant for Eq. (6-29), one obtains

$$K = \frac{[FeN_3^{2+}]_{eq}[H^+]_{eq}}{[Fe^{3+}]_{eq}[HN_3]_{eq}} = \frac{k_1 + k_2/[H^+]}{k_{-1} + k_{-2}/[H^+]} \tag{6-31}$$

This result poses a difficulty. The expression obtained suggests that K will depend upon $[H^+]$ as shown in Eq. (6-31), yet we know from thermodynamic considerations that K must be independent of $[H^+]$.

The solution to the dilemma lies in the principle of microscopic reversibility (considered more fully in Sec. 6-7), which states that at equilibrium the forward and reverse rates are the same *along each pathway*. Expressed algebraically, this statement is $k_1/k_{-1} = k_2/k_{-2}$, and substitution of that relation into Eq. (6-31) gives the result

$$K = \frac{k_1}{k_{-1}} = \frac{k_2}{k_{-2}} \qquad (6\text{-}32)$$

The discussion in this section has shown that the forward and reverse rate expressions define the thermodynamics. The inverse is not exactly true. Given the forward rate law and the thermodynamic data, the reverse rate expression may be ambiguous,[3] because the coefficients in the thermodynamic equilibrium have relative significance only (in contrast to the reaction orders in the rate law). The hypothetical first-order conversion of A to B is consistent with a reverse rate of $k_{-1}[B]$ or with the expression $k'_{-1}[B]^2/[A]$ or with the general equation $k''_{-1}[B]^n/[A]^{n-1}$. This point tends to be of limited practical importance, since chemical reactions tend to occur by simple processes, and the family of possible expressions consists largely of highly unlikely forms.

Some caution[4] is called for if one seeks to compute an equilibrium constant by division of forward and reverse rate constants. Consider the two-step mechanism of Eqs. (6-24) and (6-25) under conditions in which B is *not* a steady-state intermediate. If the rate of disappearance of A were determined very early in an experiment, before appreciable B and C have accumulated, $-d[A]/dt \cong k_1[A]$. Similar measurements starting with product would afford $-d[C]/dt = k_{-2}[C]$. Clearly, the quotient k_1/k_{-2} is not the correct value of K. The point is that the apparent rate constants, k_1 and k_{-2}, are not the values that apply when the system is at equilibrium.

6-5 CONNECTIONS BETWEEN KINETICS AND THERMODYNAMICS

A useful family of relations can be derived by comparing the relations for the temperature dependence of a rate constant with the thermodynamic expression for the equilibrium constant.

Based on the Arrhenius equation, we have

$$\frac{k_f}{k_r} = \frac{A_f \exp\left(-E_f/RT\right)}{A_r \exp\left(-E_r/RT\right)} \qquad (6\text{-}33)$$

Comparison with the thermodynamic relation

$$K = \exp\left(-\frac{\Delta G^\circ}{RT}\right) = \exp\left(\frac{\Delta S^\circ}{R}\right) \exp\left(-\frac{\Delta H^\circ}{RT}\right) \qquad (6\text{-}34)$$

leads to the equation[†]

$$E_f - E_r = \Delta H°$$ (6-35)

Similar treatment starting the the equation of the activated complex theory gives the results

$$\Delta H_f^{\ddagger} - \Delta H_r^{\ddagger} = \Delta H°$$ (6-36)

$$\Delta S_f^{\ddagger} - \Delta S_r^{\ddagger} = \Delta S°$$ (6-37)

Those results, and a restatement of Eq. (6-23),

$$\frac{k_f}{k_r} = \exp\left(-\frac{\Delta G°}{RT}\right)$$

suggest that each rate constant is capable of expression in terms of a free energy of activation $\Delta G^{\ddagger}$:

$$\ln k_f = -\frac{\Delta G^{\ddagger}}{RT} + \text{constant}$$ (6-38a)

$$\ln k_r = -\frac{\Delta G^{\ddagger}}{RT} + \text{constant}$$ (6-38b)

Also, the expressions of (6-38) for the free energy of activation become

$$\ln k_f = -\Delta G_f^{\ddagger} + \ln \frac{RT}{Nh}$$ (6-39)

$$\Delta G_f^{\ddagger} = -RT \ln \frac{k_f Nh}{RT}$$ (6-40)

$$\Delta G_f^{\ddagger} = \Delta H_f^{\ddagger} - T \Delta S_f^{\ddagger}$$ (6-41)

6-6 THE HEAT CAPACITY OF ACTIVATION

It was pointed out earlier in this chapter that curvature in the plot of $\ln (k/T)$ versus T^{-1}, which might signal a complicated mechanism, may arise from a temperature-dependent enthalpy of activation. By analogy with the familiar thermodynamic equation, the heat capacity of activation, $\Delta C_p^{\ddagger}$, is defined by the equation

$$\Delta C_p^{\ddagger} = \frac{d \Delta H^{\ddagger}}{dT}$$ (6-42)

Highly precise kinetic data giving accurate values of rate constants and temperatures are needed to obtain values of $\Delta H^{\ddagger}$ which are sufficiently precise to warrant treatment by this method. Even for reactions in which the value of $\Delta C_p^{\ddagger}$ is appreciable, the values between ±100 and ±400 J mol^{-1} K^{-1}, the extent of curvature in a plot of

[†]This equation is strictly applicable only in a condensed phase. In the gas phase, however, where the rate constants are defined in terms of concentration changes but the equilibrium constant in terms of partial pressures with $\Delta H° = -Rd \ln K_p/d(1/T)$, then $E_f - E_r = \Delta E°$.

Table 6-1 Kinetic data^a for the hydrolysis of methylchloroformate, Eq. (6-43)

Temp./°C	$10^4 \ k/s^{-1}$	$\overline{(\Delta H^{\ddagger})}/kJ \ mol^{-1}$ ^{b}	$\frac{1}{2}(T'' - T')/°C$
0.570	0.4209		
4.976	0.7016	73.50	2.77
10.031	1.229	72.60	7.50
14.987	2.087	72.53	12.51
19.569	3.323	71.09	17.28
25.025	5.642	70.42	22.30
29.832	8.812	69.68	27.43
35.012	14.05	69.51	32.44
40.133	21.72	68.72	37.59
45.135	32.65	67.51	42.63

^{a}From Ref. 5; ^{b}Calculated from adjacent points according to Eq. (6-45).

$\ln (k/T)$ versus T^{-1} is slight. Consider the hydrolysis of methylchloroformate:

$$ClC(O)OCH_3 + H_2O = CO_2 + CH_3OH + H^+ + Cl^- \tag{6-43}$$

Values of the first-order rate constant[5] are given in Table 6-1, and the plot of $\ln (k/T)$ versus T^{-1} in Fig. 6-2. The latter is barely curved on the scale of this graph. The graph

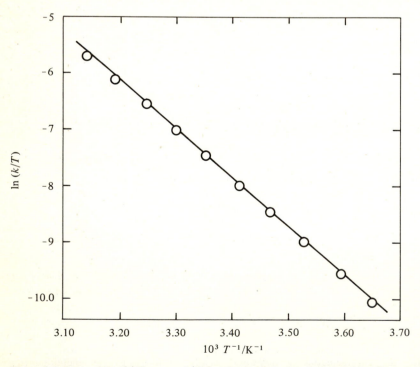

Figure 6-2 Plot of $\ln (k/T)$ versus $1/T$ for the hydrolysis of methylchloroformate (Ref. 5 and Table 6-1). The degree of curvature, although slight, is considered a real effect to which the scale of this graph cannot do justice.

hardly does justice to the accuracy of the data, however, and the extent of the curvature must be determined by other means.[5,6] The most reliable method is a numerical least-squares fit of the data to the relation

$$\ln k = A + BT^{-1} + C \ln T \qquad (6\text{-}44)$$

This fit gives parameters A, B, and C from which $\Delta H^{\ddagger}$, $\Delta S^{\ddagger}$, and $\Delta C_p^{\ddagger}$ are then calculated.[6] Unfortunately, however, the method is not very instructive of the magnitude of the effect or the physical situation. For that reason a cruder but more intuitive method has been developed for the data at hand, a method less correct but considerably easier to appreciate. Approximately, it is this: Over a sufficiently narrow range of temperature, $\Delta H^{\ddagger}$ is constant. One can approximate a series of such values by calculating a mean value of the activation enthalpy $\overline{\Delta H^{\ddagger}}$ from successive pairs of values of (k', T'), (k'', T''):

$$\overline{(\Delta H^{\ddagger})} = \frac{R \ln (k''/k')}{1/T' - 1/T''} \qquad (6\text{-}45)$$

The third column of Table 6-1 presents such values, which are seen to decrease smoothly from 73.50 to 67.51 kJ mol^{-1} with increasing temperature. According to Eq. (6-42), the average values are then plotted as $\overline{(\Delta H^{\ddagger})}$ versus T_{avg}, giving a straight line of slope $\Delta C_p^{\ddagger}$. The graph is shown in Fig. 6-3, and it is seen to define a reasonably straight line having a slope $\Delta C_p^{\ddagger} = -150$ J mol^{-1} K^{-1}, which agrees with the accepted value.

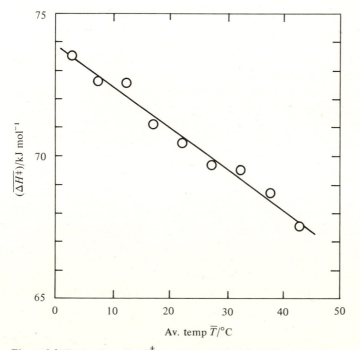

Figure 6-3 Evaluation of $\Delta C_p^{\ddagger}$ for the hydrolysis of methylchloroformate (Table 6-1 and Ref. 5). This plot shows the incremental, average enthalpy of activation from Eq. (6-45) versus the average temperature in the interval.

The value of $\Delta C_p^{\ddagger}$ may provide insight into the nature of the transition state, especially with respect to solvation and charge separation. Large values are often found for reactions leading to the formation of ions from neutral molecules. To evaluate $\Delta C_p^{\ddagger}$ requires kinetic data of unusually high accuracy. The differential temperature method of Albery and Robinson[7] can be used to provide activation parameters of high accuracy. The authors point out that if concentrations are measured to an accuracy of 1 percent at two temperatures at which the ratio of rate constants is 1.1, the uncertainty in $\Delta H^{\ddagger}$ is of the order of 20 percent. On the other hand, it is possible to obtain highly precise values of $\Delta H^{\ddagger}$ by measuring the *difference* in concentration between two reactions held at different known temperatures. For example, one might use a double-beam spectrophotometer with reactions in progress in both sample and reference compartments. The two reactions would be made up identically but thermostatted at different temperatures. The instrument recording presents, as a function of time, the concentration difference between the two, measured far more accurately than if each run were conducted independently. Results from such an experiment are considered in Prob. 6-4.

6-7 THE PRINCIPLE OF MICROSCOPIC REVERSIBILITY

The basic statement is that the forward and reverse of a particular pathway are related as mirror images. In mechanistic terms this requires that the mechanism of the reverse reaction is the same as that of the forward one (determined under comparable conditions). The positions of the atoms as the reactions move along the reaction coordinate to form and later destroy the activated complex are identical in both directions, save that the signs of the momentum vectors are reversed. Consequently, a condition of equilibrium is not just that gross forward and reverse rates are identical but that the opposing rates along each pathway are equal.

We shall consider two examples, the first being Cl^- loss from the geometrical isomers of the dichlorochromium(III) ion.[8,9] The reaction scheme is that shown in Eq. (6-46). The reaction has been studied by two methods. In the first instance,[8]

$$cis\text{-}CrCl_2^+ \underset{tc}{\overset{ct}{\rightleftharpoons}} trans\text{-}CrCl_2^+$$

$$\text{cm} \Big\Updownarrow \text{mc} \qquad \text{tm} \Big/\!\!\Big/ \text{mt} \tag{6-46}$$

$$CrCl^{2+} + Cl^-$$

samples were removed from the mixture, quenched, and the components separated chromatographically. Later, a method was developed to disentangle the rate constants from simultaneous multiwavelength spectrophotometric determinations on the unseparated mixture.[9] The latter constitutes an example of the techniques of numerical integration applied to a complex kinetic pattern.

The rate constants applicable to the reaction are given in Table 6-2. The principle of microscopic reversibility provides this relation:

$$K_{ct} = \frac{k_{ct}}{k_{tc}} = \frac{k_{mt}}{k_{tm}} \times \frac{k_{cm}}{k_{mc}} \tag{6-47}$$

Since cis-trans isomerization proceeds directly, as well as by reaction of $CrCl^{2+}$ and Cl^-, experiments starting with, say, initially pure *trans*-$CrCl_2^+$ must properly take

Table 6-2 Kinetic and thermodynamic parameters[a] for reactions of *cis*-CrCl₂⁺, *trans*-CrCl₂⁺, and CrCl²⁺

Reaction	k_f	k_r/s^{-1}	K
$CrCl^{2+} + Cl^- \rightleftarrows cis\text{-}CrCl_2^+$	$k_{mc} = 3.6 \times 10^{-7}$ M⁻¹ s⁻¹	$k_{cm} = 1.8 \times 10^{-6}$	2.0×10^{-3} M⁻¹
$CrCl^{2+} + Cl^- \rightleftarrows trans\text{-}CrCl_2^+$	$k_{mt} = 1.3 \times 10^{-7}$ M⁻¹ s⁻¹	$k_{tm} = 8.3 \times 10^{-5}$	1.6×10^{-3} M⁻¹
$cis\text{-}CrCl_2^+ \rightleftarrows trans\text{-}CrCl_2^+$	$k_{ct} = 6.2 \times 10^{-5}$ s⁻¹	$k_{tc} = 7.8 \times 10^{-5}$	0.78

[a]Refer to Eq. (6-47). Values, taken from Ref. 8, are at 34.8°C.

into account the parallel reactions of isomerization and aquation. (Particularly so, since in the absence of added Cl⁻ the reactions of CrCl²⁺ and Cl⁻ are of negligible importance.) The kinetic expressions, with such terms neglected, are

$$\frac{d[CrCl^{2+}]}{dt} = k_{tm}[\text{trans}] + k_{cm}[\text{cis}] \tag{6-48}$$

$$\frac{-d[\text{cis}]}{dt} = k_{tc}[\text{trans}] - (k_{ct} + k_{cm})[\text{cis}] \tag{6-49}$$

$$\frac{-d[\text{trans}]}{dt} = -(k_{tc} + k_{tm})[\text{trans}] + k_{ct}[\text{cis}] \tag{6-50}$$

The numerical values are such that the steady-state approximation is not valid. The experimental data are depicted in Fig. 6-4, and in the same figure are shown solid lines calculated from a numerical integration of these equations.

A second example is that given by T. L. Brown[10] for the mechanism of exchange of carbon monoxide between free CO and the metal carbonyl complex Mn(CO)₅Br. The experimental finding is that (1) reaction of ¹³CO with Mn(CO)₅Br forms Mn(CO)₄(¹³CO)Br in which (*a*) it can be unambiguously shown by infrared spectroscopy that both cis and trans positions have incorporated the ¹³C label and (*b*) the rate of exchange consists of a single term, $k_{ex}[\text{Mn(CO)}_5\text{Br}]$, independent of [CO] and (2) the ¹³CO stretching frequencies in cis and trans products grow in at the same rate.

Those observations originally suggested a mechanism rejected by Brown because it violates the principle of microscopic reversibility. The original proposal invoked these steps: rate-limiting dissociation of a *cis*-CO (the *trans*-CO was presumed to be inert), with capture of a ¹³CO by either isomer of the stereochemically labile five-coordinate intermediate. The sequence is depicted in Eq. (6-51).

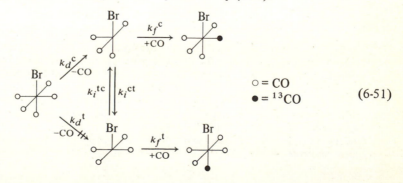

$$\tag{6-51}$$

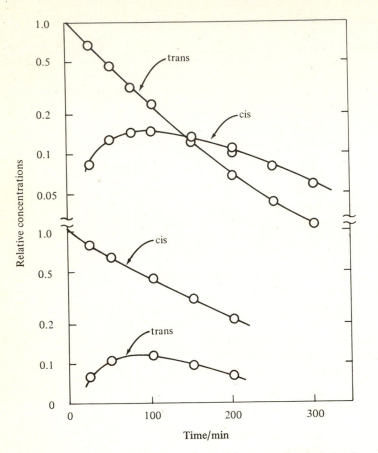

Figure 6-4 Relative concentrations (log scale) versus time, depicting [*trans*-$CrCl_2^+$] ($\circ$) and [*cis*-$CrCl_2^+$] ($\circ-$). The upper part refers to a run starting with pure trans isomer, the lower to a run starting with pure cis. The points are experimental data,[6] and lines are those calculated from the rate constants in Table 6-1.

Brown points out that such a mechanism violates the principle of microscopic reversibility in that the relative rates of capture of CO must be in proportion to their dissociation, $k_f^c/k_f^t = k_d^c/k_d^t$. The resolution in this case originates because the exchange reaction necessarily has a perfectly symmetrical reaction coordinate, which the proposal in (6-52) disregards. Brown states:

> If there is preferential loss of CO from the equatorial position, then, no matter what modes of equilibration may prevail in the five-coordinate intermediate, addition of CO must occur at the equatorial position, with the same degree of preference which characterizes loss of CO. This conclusion follows directly from the principle of microscopic reversibility. If the transition state for loss of axial CO possesses a higher free energy than that for loss of radial CO, the transition state for addition of CO to the intermediate along the axis must be correspondingly higher. The key to the application of this concept in the present context is that the labeling which is observed resides in the group being added to the intermediate and not in the parent molecule.

The correct application of this principle has proved of occasional difficulty; particular aspects are discussed more fully in the literature.[11,12]

REFERENCES

1. Treiling, E. C., H. C. Johnston, and R. A. Ogg, Jr.: *J. Chem. Phys.*, **20**:327 (1952).
2. (a) Seewald, D., and N. Sutin: *Inorg. Chem.*, **2**:643 (1963); (b) D. W. Carlyle and J. H. Espenson, *Inorg. Chem.*, **6**:1370 (1967).
3. Manes, M., L. J. E. Hofer, and S. Weller: *J. Chem. Phys.*, **22**:1612 (1954).
4. Laidler, K. J.: "Chemical Kinetics," 2d ed., McGraw-Hill Book Company, New York, 1965, p. 69.
5. Queen, A.: *Can J. Chem.*, **45**:1612 (1967).
6. Kohnstan, G.: *Advances in Phys. Org. Chem.*, **5**:121 (1967).
7. Albery, W. J., and B. H. Robinson: *J. Chem. Phys.*, **20**:327 (1952).
8. Salzman, J. D., and E. L. King: *Inorg. Chem.*, **6**:426 (1967).
9. Mønsted, L., and O. Mønsted, *Acta Chem. Scand.*, **32**:19 (1978).
10. Brown, T. L.: *Inorg. Chem.*, **7**:2673 (1968).
11. Burwell, R. L., Jr., and R. G. Pearson: *J. Chem. Phys.*, **70**:300 (1969).
12. Abraham, M. H., D. Dodd, M. D. Johnson, E. S. Lewis, and R. A. More O'Farrell: *J. Chem. Soc., B*, 762 (1971).

PROBLEMS

6-1 *Temperature dependence.* The reaction in Prob. 3-1 was also studied under the same conditions at 35.0°C. Use the data below to evaluate (*a*) K at 35.0°C, (*b*) k_f and k_r at 35.0°C, (*c*) $\Delta H^{\ddagger}$ and $\Delta S^{\ddagger}$ for forward and reverse rate constants, (*d*) ΔH° and ΔS°.

Time/s	0.0	600	1200	1800	3600	18,000
Absorbance 285 nm, 35.0°C	0.450	0.573	0.655	0.708	0.783	0.814

6-2 *Forward and reverse reaction rates.* The rate of isotopic exchange between $U^{4+}(aq)$ and UO_2^{2+} in dilute perchloric acid is given by (see Prob. 5-14):

$$R_{ex} = k_{ex} \frac{[U^{4+}][UO_2^{2+}]}{[H^+]^3}$$

with $k = 2.1 \times 10^{-7}$ M^2 s^{-1} at 25°C. The equilibrium constant is 2×10^{-10} M^4 (25°C) for the reaction

$$U^{4+} + UO_2^{2+} + 2H_2O = 2UO_2^+ + 4H^+$$

Assuming that the latter reaction constitutes the major pathway for U(IV)-U(VI) exchange, determine the form of the rate law for U(V) disproportionation and its rate constant.

6-3 *Kinetics and thermodynamics; microscopic reversibility.* The elementary gas-phase reaction

$$NO + ClNO_2 = NO_2 + ClNO$$

is characterized by these thermodynamic parameters at 298 K: $\Delta H^{\circ} = -16.5$ kJ mol^{-1} and $\Delta G^{\circ} = -22.9$ kJ mol^{-1}. The forward rate constant can be expressed in Arrhenius form [Eq. (6-9)] and by an activated complex theory expression [Eq. (6-13)]. (In both cases use 1 mol/dm^3 standard state.)

(a) Express k_r in the same two forms. Compute k_r at 298 K.

(b) What experiments using nitrogen-15-labeled compounds might distinguish the molecular mechanisms of chlorine atom and oxygen atom transfer?

6-4 *Differential temperature method.* The method described in the text has been applied[7] to a study of the iodination of acetone, a pseudo-zero-order reaction under the conditions employed ($[(CH_3)_2CO] \gg [I_2]$)

$$\frac{-d[I_2]}{dt} = k[(CH_3)_2CO] = k'$$

Two identical reaction solutions were prepared, one at the lower temperature T_1 in the sample compartment of a double-beam spectrophotometer, the other at T_2 in the reference beam. A direct recording of $\Delta D = D_1 - D_2$ was made as a function of time while the difference in reaction temperature was maintained to ±0.001°C.

Evaluate k_2'/k_1' and $\Delta H^{\ddagger}$ for the run shown.

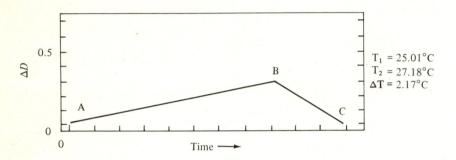

6-5 *Composite temperature dependence.* Consider the two mechanisms shown, each of which leads to an expression $-d[A]/dt = k_{app}[A]$. In each case show the shape of a plot of $\ln k_{app}$ versus $1/T$.

Mechanism 1:

$$A \begin{array}{c} \xrightarrow{1} B + C \\ \searrow_2 A^* \xrightarrow{Fast} B + C \end{array}$$

Mechanism 2:

$$A \underset{-3}{\overset{3}{\rightleftarrows}} A^*$$

$$*A \xrightarrow{4} B + C$$

6-6 *Kinetics and equilibrium.* Rabinovitch and Michel [*J. Am. Chem. Soc.*, 81:5065 (1959)] report that the cis-trans isomerization of 2-butene follows first-order kinetics at high pressures. The value of k_f is 1.9×10^{-5} s^{-1} at 742°K and has $E_a = 2.67 \times 10^4$ J mol^{-1}. The equilibrium constant is 1.53 and has $\Delta H° = 0$. Consider an experiment with $[cis]_0 = 7.16 \times 10^{-4}$ mol dm^{-3} and $[trans]_0$. Compute (a) the apparent rate constant for approach to equilibrium at 742 K, (b) [cis] at time 6×10^4 s, and (c) the apparent rate constant at 686 K.

6-7 *Composite temperature dependence.* A. Burg [*J. Am. Chem. Soc.*, 74:3482 (1955)] and M. E. Garabadian and S. W. Benson [*J. Am. Chem. Soc.*, 86:186 (1964)] have found that the rate of the gas-phase reaction $2BH_3CO = B_2H_6 + 2CO$ is

$$\frac{-d[BH_3CO]}{dt} = k_{exp} \frac{[BH_3CO]^2}{[CO]}$$

with k_{exp}, in Arrhenius form, given by

$$k_{exp}/s^{-1} = 7.6 \times 10^{13} \exp\left(-\frac{111{,}900 \text{ J mol}^{-1}}{RT}\right)$$

Chien and Bauer [*Inorg. Chem.*, **16**:867 (1977)] give thermodynamic data at 298 K (converted to a standard state of 1 mol/cm^3):

	BH$_3$CO	BH$_3$	CO
ΔH_f°/kJ mol^{-1}	−111.0	+99.9	−111.5
S°/J mol^{-1} K^{-1}	+165.7	+103.8	+113.5

This reaction involves BH$_3$ as a steady-state intermediate reacting in this rate-limiting step:

$$\text{BH}_3\text{CO} + \text{BH}_3 \xrightarrow{\ 2\ } \text{B}_2\text{H}_6 + \text{CO}$$

Evaluate k_2 in an Arrhenius form.

6-8 *Temperature dependence.* Derive a general relation between $\Delta S^\ddagger$ of activated complex theory and A of the Arrhenius equation. Compute A for a first-order reaction with $\Delta S^\ddagger = -20$ J mol^{-1} K^{-1} at 298 K.

6-9 *Elementary reactions.* Castellano and Schumacher [*Z. Phys. Chem.*, **24**:198 (1962)] considered the photochemical decomposition of ozone:

$$\text{O}_3 + h\upsilon \longrightarrow \text{O}_2 + \text{O}$$

The O atoms so produced either reacted to form O$_2$ or returned to ozone in the presence of a third body M:

$$\text{O} + \text{O}_3 \xrightarrow{\ 1\ } 2\text{O}_2$$

$$\text{O} + \text{O}_2 + \text{M} \xrightarrow{\ 2\ } \text{O}_3 + \text{M}$$

The ratios of rate constants are as follows for different M's.

M	Ar	He	N$_2$	O$_2$	CO$_2$	O$_3$
$k_1 k_2^{-1}$/mol dm^{-3}	550	420	370	310	150	140

Consider now the thermal dissociation of O$_3$:

$$\text{O}_3 + \text{M} \xrightarrow{\ 3\ } \text{O} + \text{O}_2 + \text{M}$$

Compute the relative values of k_3 for different M's.

6-10 *Composite temperature dependence.* Show that a plot of the apparent first-order rate constant of Eq. (6-22) as $\ln (k_{app}/T)$ versus $1/T$ will always be curved downward, corresponding to an apparent $\Delta H^\ddagger$ which decreases with increasing temperature. What will the shape of the plot be if, by chance, $\Delta H_6^\ddagger = \Delta H_{-5}^\ddagger$?

SEVEN

CHAIN REACTIONS

Previous consideration has been directed toward reactions in which products form by a limited succession of elementary steps or by parallel sequences of such steps. The over-all reaction is completed when products appear from the specified quantity of reactants in accord with the net equation.

A *chain reaction sequence* is one which the key intermediates, often atomic or radical species, may be generated in one step and reenter the reaction sequence at another.

7-1 CHARACTERISTICS OF CHAIN REACTIONS

We recognize two classes of chain reactions, one in which the rate law correctly gives the stoichiometric composition of the transition state and another in which it does not. An example of the latter is the thermal decomposition of acetaldehyde (Sec. 7-2), which proceeds at a rate[1] given by

$$\frac{-d[CH_3CHO]}{dt} = k[CH_3CHO]^{3/2} \tag{7-1}$$

Clearly, this expression cannot be directly interpreted to yield the elemental composition of the activated complex.

A radical chain reaction consists of three main reaction types, termed initiation, propagation, and termination. The *initiation process* may be a unimolecular or bimolecular step, and quite often it is a photoinitiated reaction. It serves to generate a reactive intermediate, but the step itself does not usually provide a route for the consumption of a substantial part of one of the reactants.

Chain-propagating or *chain-carrying steps* are often bimolecular reactions of one intermediate and one substrate, and each has as a product the intermediate used in the other. In the simplest mechanisms there are two such steps, and they are the reactions which generally accomplish the bulk of the chemical reaction. A situation frequently encountered is one in which the rates of the chain-propagating steps are much greater than the rate of the initiation and termination reaction (the latter to be discussed shortly). In such cases, it is said that the chains are long or the chain length is high. Chain length is defined as the ratio of the overall rate to the rate of initiation, and the overall rate is essentially just the rate of a propagation step. Qualitatively that means the propagating steps accomplish virtually all of the reaction. In such cases the sum of the propagation steps gives the overall net stoichiometric reaction.

The recycling of the two chain-carrying intermediates can be interrupted by a *termination reaction*. The process converts one of the active chain-carrying intermediates to stable molecules such as reactants, products, or inert by-products, or perhaps to chemically ineffective radicals. This step does not usually constitute a major source of products, provided the chains are long. The particular termination step does exert a major influence, however, upon the mathematical form of the rate expression.

Further general comments about rate laws and steady-state approximations in chain reactions are given in Sec. 7-6, but it is instructive at this point to consider several illustrative examples.

7-2 THE DECOMPOSITION OF ACETALDEHYDE

The main reaction occurring in the gas phase is

$$CH_3CHO = CH_4 + CO \tag{7-2}$$

Trace quantities of C_2H_6 and H_2 are also detected in the products. The $\frac{3}{2}$-order rate law was given in Eq. (7-1). Consider the following chain mechanism

Initiation: $$CH_3CHO \xrightarrow{1} CH_3\cdot + \cdot CHO \tag{7-3}$$

Propagation: $$CH_3\cdot + CH_3CHO \xrightarrow{2} CH_4 + CH_3\cdot CO \tag{7-4}$$

$$CH_3\cdot CO \xrightarrow{3} CH_3\cdot + CO \tag{7-5}$$

Termination: $$2CH_3\cdot \xrightarrow{4} C_2H_6 \tag{7-6}$$

The rate law can be derived by making the steady-state approximation for the chain-carrying radicals:

$$\frac{d[CH_3\cdot]}{dt} = k_1[CH_3CHO] - k_2[CH_3\cdot][CH_3CHO] + k_3[CH_3\cdot CO] - 2k_4[CH_3\cdot]^2 = 0 \tag{7-7}$$

$$\frac{d[CH_3\cdot CO]}{dt} = k_2[CH_3\cdot][CH_3CHO] - k_3[CH_3\cdot CO] = 0 \tag{7-8}$$

Addition gives the result

$$k_1[CH_3CHO] - 2k_4[CH_3\cdot]^2 = 0 \tag{7-9}$$

$$[CH_3\cdot]_{ss} = \sqrt{\frac{k_1[CH_3CHO]}{2k_4}} \tag{7-10}$$

from which

$$\frac{-d[CH_3CHO]}{dt} = k_1[CH_3CHO] + k_2[CH_3CHO][CH_3\cdot]_{ss} \tag{7-11}$$

$$\frac{-d[CH_3CHO]}{dt} = k_1[CH_3CHO] + k_2\left(\frac{k_1}{2k_4}\right)^{1/2}[CH_3CHO]^{3/2} \tag{7-12}$$

This result agrees with the experimental $\frac{3}{2}$-order expression provided the second term is much larger than the first. It is referred to as the approximation of "long chains," and it amounts to saying that the propagation steps consume practically all of the acetaldehyde.

The experimental rate constant k of Eq. (7-1) is given by Eq. (7-13). Its Arrhenius activation energy is 201 kJ mol^{-1}, which is considerably less than the thermodynamic

$$k = k_2\left(\frac{k_1}{2k_4}\right)^{1/2} \tag{7-13}$$

value for the homolysis of acetaldehyde, $\Delta H_1^\circ = 345$ kJ mol^{-1}.

The experimental activation energy[1] is, according to Eq. (7-13)

$$E_a = E_2 + \tfrac{1}{2}E_1 - \tfrac{1}{2}E_4 = 201 \text{ kJ mol}^{-1}$$

The value of E_2 has been measured independently,[2] and it is 31.4 kJ mol^{-1}. Taking $E_4 \simeq 0$ (as usual for radical recombinations) and taking $E_{-1} \simeq 0$ (another radical recombination), we have $E_1 \simeq \Delta H^\circ - RT$ (since $\Delta E^\circ = \Delta H^\circ - \Delta nRT$ for gas-phase reactions). A calculated value of E_a is thus $31.4 + \tfrac{1}{2}(345 - RT) \sim 200$ kJ mol^{-1}, which is in good agreement with the experimental value.

The formyl radical produced in the initiation reaction of Eq. (7-3) does not enter the chain reaction. Ultimately it must produce stable products, and it is thought that the following sequence constitutes a secondary chain process:

$$\cdot CHO + M \xrightarrow{5} CO + H\cdot + M^\dagger \tag{7-14}$$

$$H\cdot + CH_3CHO \xrightarrow{6} H_2 + CH_3\cdot CO \tag{7-15}$$

But because H_2 is a minor product, $<1\%$ of CH_4, this chain is far less important.

The qualitative phrase *chain length* has been used. We may define it as

$$\text{Chain length} = \frac{\text{rate of appearance of products}}{\text{rate of radical formation by initiation}} \tag{7-16}$$

---

†M stands for a third body, such as bulk molecules, possibly walls. Reactions involving atoms need to recombine in the presence of a third body for momentum to be conserved.

Applied to the reaction at hand, this is

$$\text{Chain length} = \frac{k_2(k_1/2k_4)^{1/2}[CH_3CHO]^{3/2}}{k_1[CH_3CHO]} = \frac{k_2}{\sqrt{k_1k_4}}[CH_3CHO]^{1/2}$$

By inserting values at 800 K, $k = 5.4 \times 10^{-3}$ dm$^{3/2}$ mol$^{-1/2}$ s^{-1} and $k_1 \approx 1.8 \times 10^{-7}$ s^{-1} as given by Benson,[3] the chain length is found to be $3 \times 10^4 [CH_3CHO]^{1/2}$. Even at concentrations as low as 10^{-4} mol dm^{-3} (which corresponds to a pressure of 665 Pa or 5 Torr at 800 K) the chain length is 300 and the long-chain approximation can be seen to be valid.

7-3 THE HYDROGEN-BROMINE REACTION

The experimental rate of reaction is given by

$$\frac{d[HBr]}{dt} = \frac{A[H_2][Br_2]^{3/2}}{[Br_2] + B[HBr]} \tag{7-17}$$

with A (in collision theory form) given by

$$A/cm^{3/2} \text{ mol}^{-1/2} \text{ s}^{-1} = 1.43 \times 10^{13} \, T^{1/2} \exp\left(-\frac{170,100 \text{ J mol}^{-1}}{RT}\right) \tag{7-18}$$

$B = 0.12$ mol cm^{-3} (roughly independent of T: 0.116 at 298 K, 0.122 at 573 K).

The mechanism to account for these results is given, along with thermodynamic data, as follows:

Reaction	$\Delta H°/\text{kJ mol}^{-1}$	
$Br_2 + M \underset{-1}{\overset{1}{\rightleftharpoons}} 2Br\cdot + M$	+192.8	(7-19)
$Br\cdot + H_2 \underset{-2}{\overset{2}{\rightleftharpoons}} HBr + H\cdot$	+ 69.5	(7-20)
$\underline{H\cdot + Br_2 \overset{3}{\longrightarrow} HBr + Br\cdot \quad -173.1}$		(7-21)
Net: $Br_2 + H_2 = 2HBr$	-103.6	(7-22)

In this particular instance, the initiation and termination steps are the reverse of one another. Making the steady-state approximation for $[Br\cdot]$ and $[H\cdot]$,

$$\frac{d[Br\cdot]}{dt} = 0 = 2k_1[Br_2][M] - 2k_{-1}[Br\cdot]^2[M] - k_2[Br\cdot][H_2]$$

$$+ k_{-2}[HBr][H\cdot] + k_3[H\cdot][Br_2] \tag{7-23}$$

$$\frac{d[H\cdot]}{dt} = 0 = k_2[Br\cdot][H_2] - k_{-2}[HBr][H\cdot] - k_3[H\cdot][Br_2] \tag{7-24}$$

Combining the two equations gives

$$[Br\cdot]_{ss} = \left(\frac{k_1[Br_2]}{k_{-1}}\right)^{1/2} \tag{7-25}$$

$$[H\cdot]_{ss} = \frac{K_1^{1/2}k_2[Br_2]^{1/2}[H_2]}{k_{-2}[HBr] + k_3[Br_2]} \tag{7-26}$$

The reaction rate is

$$\frac{d[HBr]}{dt} = k_2[Br\cdot][H_2] - k_{-2}[HBr][H\cdot] + k_3[H\cdot][Br_2] \tag{7-27}$$

$$\frac{d[HBr]}{dt} = 2k_3[H\cdot][Br_2] = \frac{2k_2k_3(k_1/k_{-1})^{1/2}[Br_2]^{3/2}[H_2]}{k_{-2}[HBr] + k_3[Br_2]} \tag{7-28}$$

Comparison with the experimental equation identifies parameters A and B of Eq. (7-17) as

$$A = 2k_2\left(\frac{k_1}{k_{-1}}\right)^{1/2} \tag{7-29}$$

$$B = \frac{k_{-2}}{k_3} \tag{7-30}$$

The (originally) puzzling question, how can dissociation of Br_2 be involved given that $D(Br_2) = 193$ kJ mol^{-1}, whereas E_a is only 170 kJ mol^{-1}, is answered. The expression for E_a of parameter A is

$$E_A = \tfrac{1}{2}\Delta H_1^\circ + E_2 \tag{7-31}$$

according to which

$$E_2 = E_A - \tfrac{1}{2}\Delta H_1^\circ = 170.1 - \tfrac{1}{2}(192.8) = 73.7 \text{ kJ mol}^{-1} \tag{7-32}$$

Combining the thermodynamic expression for Eq. (7-19), which at a standard state of 1 mol cm^{-3} is

$$K_1/\text{mol cm}^{-3} = \exp\left(\frac{21.2 \text{ J mol}^{-1} \text{ K}^{-1}}{R}\right) \exp\left(-\frac{192,800 \text{ J mol}^{-1}}{RT}\right) \tag{7-33}$$

with Eq. (7-18) gives

$$k_2/\text{cm}^3 \text{ mol}^{-1} \text{ s}^{-1} = 2.0 \times 10^{12} \, T^{1/2} \exp\left(-\frac{73,700 \text{ J mol}^{-1}}{RT}\right) \tag{7-34}$$

Also, since B is independent of T, $E_{-2} \sim E_3$. But ΔH° for Eq. (7-20) is 69.5 kJ/mol, and that value in conjunction with Eq. (7-35) gives $E_{-2} \sim E_3 \sim 4.2$ kJ mol^{-1}.

$$\Delta H_{(7\text{-}20)} = E_2 - E_{-2} \tag{7-35}$$

One notes that the steady-state concentration of atomic bromine is, according to Eq. (7-25), simply the concentration in thermal equilibrium with molecular bromine. The reason is that the initiation and termination reactions, which establish the concentration, are the forward and reverse of the dissociation process. In contrast, the steady-state concentration of H· considerably exceeds the value in thermal equilibrium with H_2 (see Prob. 7-10).

Several conceivable reactions were omitted from the mechanism, including those given in Eqs. (7-36) and (7-38):

Reaction	ΔH°/kJ mol^{-1}	
$H_2 + Br_2 \longrightarrow 2HBr$	-103.3	(7-36)
$H_2 + M \longrightarrow 2H\cdot + M$	$+436.0$	(7-37)
$H\cdot + Br\cdot + M \longrightarrow HBr + M$	-366.1	(7-38)

In general, these reactions are rejected because their inclusion would lead to a different kinetic expression. Also, formation of H· by direct dissociation of H_2 [Eq. (7-37)] would require too high an activation energy. The alternative termination step [Eq. (7-38)] is far less important than bromine atom recombination because H· reacts so efficiently with Br_2 and $[Br_2] \gg [Br\cdot]_{ss}$. Additionally, the reverse of reaction (7-21) is unimportant owing to the high activation energy: $E_{-3} = 177$ kJ mol^{-1}. A thorough analysis of these "missing reactions" is given by Benson.[4]

In contrast to the reaction of acetaldehyde, that of H_2 and Br_2 is a "well-behaved" chain reaction. The reader will note that application of the concepts introduced in Chap. 5 to the experimental rate law of Eq. (7-17) correctly yields the composition of the transition states as $[BrH_2]^\ddagger$ and $[HBr_2]^\ddagger$, which correspond to the two propagation steps.

7-4 THE PHOTOCHEMICAL HYDROGEN-BROMINE REACTION

An initiation step alternative to thermal dissociation of bromine is that from bromine photodissociation. The experimental rate of the photochemical reaction is a function of the total pressure P and the intensity of absorbed light; it is given by

$$\frac{d[HBr]}{dt} = \frac{k(I_{abs})^{1/2}[H_2][Br_2]}{P^{1/2}([Br_2] + k'[HBr])} \tag{7-39}$$

Assuming the same propagating and termination reactions, the rate law can be accounted for as follows. The rate of production of Br atoms by photoinitiation is $2I_{abs}$. Steady-state relations now yield

$$2I_{abs} = 2k_{-1}[Br\cdot]^2[M] \tag{7-40}$$

$$[Br\cdot]_{ss} = \left(\frac{I_{abs}}{k_{-1}[M]}\right)^{1/2} \tag{7-41}$$

Substituting into $d[HBr]/dt$ gives

$$\frac{d[HBr]}{dt} = \frac{2(I_{abs}/k_{-1})^{1/2}k_2[H_2][Br_2]}{[M]^{1/2}\{[Br_2] + (k_{-2}/k_3)[HBr]\}} \tag{7-42}$$

Since [M] is proportional to the total pressure, the derived expression is of the same form as that found experimentally.

7-5 FREE-RADICAL HALOGENATIONS OF HYDROCARBONS

Hydrocarbons undergo free-radical halogenation by a chain mechanism involving any of three possible termination steps:

Initiation:
$$X_2 + M \xrightarrow{\ 1\ } 2X\cdot + M \tag{7-43}$$

Propagation:
$$X\cdot + RH \xrightarrow{\ 2\ } R\cdot + HX \tag{7-44}$$

$$R\cdot + X_2 \xrightarrow{\ 3\ } RX + X\cdot \tag{7-45}$$

Termination:
$$2X\cdot + M \xrightarrow{\ k_{XX}\ } X_2 + M \tag{7-46}$$

$$X\cdot + R\cdot \xrightarrow{\ k_{XR}\ } RX \tag{7-47}$$

$$R\cdot + R\cdot \xrightarrow{\ k_{RR}\ } R_2 \tag{7-48}$$

The general expression involving all three possible termination reactions is quite complex. In some cases, two competing termination reactions must be considered. For the purpose at hand, we consider the three possible termination steps separately. The usual steady-state approximations for [R·] and [X·] give these rate expressions.

Termination step	Rate law, $-d[RX]/dt$	
(7-46)	$k_2 \left(\dfrac{k_1}{k_{XX}}\right)^{1/2} [X_2]^{1/2}[RH]$	(7-49)
(7-47)	$\left(\dfrac{k_1 k_2 k_3}{k_{XR}}\right)^{1/2} [X_2][RH]^{1/2}[M]^{1/2}$	(7-50)
(7-48)	$k_3 \left(\dfrac{k_1}{k_{RR}}\right)^{1/2} [X_2]^{3/2}[M]^{1/2}$	(7-51)

This first makes the point that the choice of termination step is one of the main factors deciding the form of the rate expression. In the process of trying to deduce mechanisms consistent with a given rate law, various possible termination processes should be considered to achieve consistency.

Second, of the three expressions only the one in which Eq. (7-46) is the termination process gives a kinetic expression in which the rate law correctly indicates the composition of the activated complex $[RHX]^{\ddagger}$. Thus the chain reaction is "well-behaved." Note that initiation and termination are the reverse of one another, and particularly the consequences of writing the rate expression in terms of activated complex theory (Sec. 8-3). Activated complex theory gives the following expressions:

Equation (7-46) termination:

$$\frac{d[RX]}{dt} = \frac{RT}{Nh} \left(\frac{[X_2]}{q_{X_2}}\right)^{1/2} \left(\frac{[RH]}{q_{RH}}\right) q_2^{\ddagger} \exp\left(-\tfrac{1}{2}\Delta H_1^{\circ} + E_2\right) \tag{7-52}$$

where q's represent partition functions. The *well-behaved chain reaction is characterized by a rate dependence upon the partition function of one transition state only*, whose composition may therefore be inferred from the stoichiometry of the experimental rate law.

In contrast, when either of the other rate laws is considered, the empirical reaction orders [$2X + \tfrac{1}{2}M + \tfrac{1}{2}RH$ for Eq. (7-47) and $3X + \tfrac{1}{2}M$ for (7-48)] are not simply diagnostic of mechanism. Activated complex theory expressions for Eqs. (7-50) and (7-51) are as follows:

Equation (7-47) termination:

$$\frac{d[RX]}{dt} = \frac{RT}{Nh} \left(\frac{[X_2]}{q_{X_2}}\right) \left(\frac{[RX]}{q_{RH}}\right)^{1/2} \left(\frac{[M]}{q_M}\right)^{1/2} \left(\frac{q_1^{\ddagger} q_2^{\ddagger} q_3^{\ddagger}}{q_{XR}^{\ddagger}}\right)^{1/2}$$

$$\times \exp\left(-\frac{1}{2} \frac{E_1 + E_2 + E_3 - E_7}{RT}\right) \tag{7-53}$$

Equation (7-48) termination:

$$\frac{d[RX]}{dt} = \frac{RT}{Nh} \left(\frac{[X_2]}{q_{X_2}}\right)^{1/2} \left(\frac{[M]}{q_M}\right)^{1/2} q_3^{\ddagger} \left(\frac{q_1^{\ddagger}}{q_{RR}^{\ddagger}}\right)^{1/2}$$

$$\times \exp\left(-\frac{E_3 + \tfrac{1}{2}E_1 - \tfrac{1}{2}E_{RR}}{RT}\right) \tag{7-54}$$

In these instances the rate is dependent upon the partition functions for four [step (7-47) termination] or three [step (7-48) termination] activated complexes, and they are "ill-behaved" in the sense in which that term is being used here.

It is instructive to examine the factors, aside from the intrinsic rate constant values, which control whether a single termination step will predominate and which it will be. Consider the rate expression that will result if all three of the possible termination steps are included. The steady-state approximation for $[X\cdot]$ and $[R\cdot]$ gives

$$\frac{d[X\cdot]}{dt} = 0 = 2k_1[X_2][M] - k_2[X\cdot][RH] + k_3[R\cdot][X_2]$$

$$- 2k_{XX}[X\cdot]^2[M] - k_{XR}[X\cdot][R\cdot] \tag{7-55}$$

$$\frac{d[R\cdot]}{dt} = 0 = k_2[X\cdot][RH] - k_3[R\cdot][X_2] - k_{XR}[X\cdot][R\cdot]$$

$$- 2k_{RR}[R\cdot]^2 \tag{7-56}$$

These equations and the long-chain approximation give the net reaction rate as

$$\frac{d[RX]}{dt} = k_2[RH] \times \left(\frac{k_1[X_2][M]}{k_{XX}[M] + k_{XR}(k_2[RH]/k_3[X_2]) + k_{RR}(k_2[RH]/k_3[X_2])^2}\right)^{1/2}$$

$$\tag{7-57}$$

It is the relative importance of the denominator terms which decides the form to which the general expression will simplify. Those terms are a measure of the relative steady-state concentrations of the intermediates which are present in a ratio given by

$$\frac{[R\cdot]_{ss}}{[X\cdot]_{ss}} = \frac{k_2[RH]}{k_3[X_2]} \tag{7-58}$$

Among other conclusions, it can be seen that the larger the value of the rate constant ratio k_2/k_3, other factors being equal, the higher will be $[R\cdot]_{ss}$ and the more likely termination will occur by Eq. (7-48).

7-6 STEADY-STATE APPROXIMATION IN CHAIN REACTIONS

The discerning reader will note that certain simplifications can be made in the kinetic derivations. The steady-state approximation is always equivalent to stating

$$\text{Rate of initiation} = \text{rate of termination} \tag{7-59}$$

Several of the mechanisms previously considered have arrived at that result less directly; examine (7-9), (7-25), and (7-41). The second useful relation is that the rates of the two chain-propagating steps are equal (provided the chains are long). As illustrations, consider Eqs. (7-8) and (7-24). These simplifications should not be blindly applied without considering the specific case at hand, but they can greatly reduce the required algebraic manipulations when they are applicable.

7-7 BRANCHING CHAIN REACTIONS

Branching chain reactions are a subject of considerable importance in considering flames and explosions, but the present treatment will be limited to some elementary considerations. One very poorly understood reaction is one of the first learned by a beginning student of chemistry:

$$2H_2 + O_2 = 2H_2O$$

The following are thought to be important chain-carrying steps:

$$HO\cdot + H_2 \longrightarrow H_2O + H\cdot$$
$$H\cdot + O_2 \longrightarrow HO\cdot + O\cdot$$
$$O\cdot + H_2 \longrightarrow HO\cdot + H\cdot$$

Although the first of these is a perfectly normal propagation step, the other two are unusual in that *one* radical intermediate is converted into *two*. They are referred to as *branching* reactions. Consequently, just as in nuclear chain reactions, the concentration of chain carriers can build up exponentially and lead to explosions.

At very low pressures ($\sim$2 torr at 550°C), H_2 and O_2 react smoothly and slowly; the radicals escape to the surface. At a certain critical pressure p_1 (the "first explosion limit," roughly a few torr depending upon temperature and the surface/volume ratio of the reaction chamber), the rate increases very sharply, and at no pressure up until about 100 torr can a steady rate be seen.

On the other hand, if the reaction is begun at a pressure p_2 (~200 torr, the "second explosion limit"), a steady reaction is again seen at least below 600°C. The pressure p_2 corresponds to a sufficiently high free-radical concentration such that bimolecular recombination reactions become feasible and prevent uncontrolled, exponential reaction.

At still higher pressures, now typically 2 atm, mixtures of H_2 and O_2 are again explosive above 500°C. This "third explosion limit" corresponds to a thermal explosion: the rate is now high enough that the system cannot be held at constant temperature, and explosion accompanies the rapid rise in temperature.

7-8 OSCILLATING REACTIONS

A mechanical system, such as a pendulum, can oscillate around a position of final equilibrium. It is generally accepted that chemical reactions cannot overshoot the equilibrium point. There is, however, the occasional system which does in fact show the striking phenomenon of oscillatory behavior, essentially because there exists a dual set of solutions to the steady-state equations. No sooner is one set established than the resulting concentration changes cause the system to pass to the other, and vice versa; in effect a "chemical feedback loop" is established. What that amounts to is oscillation between the two steady-state situations. No law of thermodynamics is violated, and at no point in the process is $\Delta G > 0$.

The general phenomenon is reviewed by Noyes and Field,[5] who give reference to original work and exact formulations. The treatment here given in terms of one striking example† of an oscillating reaction, the oxidation of malonic acid by bromate ions catalyzed by cerium(IV). The stoichiometry of the net reaction with excess malonic acid [$CH_2(COOH)_2 = MA$] is

$$2BrO_3^- + 3MA + 2H^+ = 2BrMA + 3CO_2 + 4H_2O \qquad (7\text{-}60)$$

The series of elementary reactions§ can be broken down into three groups. The equations are numbered as by Noyes and Field to facilitate use of Ref. 5.

Sequence α, reduction of BrO_3^- initiated by Br^-:

(1) $$BrO_3^- + Br^- \xrightarrow{2H^+} HBrO_2 + HOBr$$

†For demonstration purposes, add to a 1-dm^3 beaker 600 cm^3 water, 60 cm^3 concentrated sulfuric acid, 20 g malonic acid, 7.8 g potassium bromate, 0.7 to 0.8 g $(NH_4)_2Ce(NO_3)_6$, and sufficient "ferroin indicator," 0.025 M $Fe(phen)_3SO_4$, to give a visible coloration, typically 1 cm^3 or more. Stir magnetically; a short but variable length of time can be expected before oscillations begin.

§For purposes of simplicity H^+ is not included as a specific participant in these reactions. In writing the rate laws, [H^+] is omitted even though most, if not all, of these reactions show a dependence upon [H^+]. In this particular reaction [H^+] is usually high and constant in any run.

(2) $$HBrO_2 + Br^- \xrightarrow{H^+} 2HOBr$$

(x) $$HOBr + MA \xrightleftharpoons{Fast} BrMA + H_2O$$

Net of α: $\quad BrO_3^- + 2Br^- + 3MA + 3H^+ = 3BrMA + H_2O \qquad (7\text{-}61)$

Sequence β, catalyzed production of HOBr:

(3a) $$BrO_3^- + HBrO_2 \xrightarrow{H^+} 2BrO_2 + H_2O$$

(3b) $$BrO_2 + Ce(III) \xrightarrow{H^+} HBrO_2 + Ce(IV)$$

(4) $$2HBrO_2 \xrightarrow{-H^+} BrO_3^- + HOBr$$

Net of β: $\quad BrO_3^- + 4Ce(III) + 5H^+ = HOBr + 4Ce(IV) + 2H_2O \qquad (7\text{-}62)$

Sequence γ, reduction of Ce(IV) and production of CO_2:

$(5a)^\dagger \quad BrMA + 4Ce(IV) + 2H_2O \xrightarrow{-5H^+} Br^- + HCOOH + 2CO_2 + 4Ce(III)$

$(5b) \quad HOBr + HCOOH \xrightleftharpoons{Fast} Br^- + CO_2 + H^+ + H_2O$

Net of γ: $BrMA + 4Ce(IV) + HOBr + H_2O = 2Br^- + 3CO_2 + 4Ce(III) + 6H^+ \quad (7\text{-}63)$

The composite of the three processes α, β, and γ gives the overall stoichiometric reaction, Eq. (7-60). The steady-state approximations for the intermediates $HBrO_2$ and BrO_2 are written in simplified form $(R_1 = k_1[BrO_3^-][Br^-]$, etc.):

$$\frac{d[HBrO_2]}{dt} = 0 = R_1 - R_2 - R_{3a} + R_{3b} - 2R_4 \qquad (7\text{-}64)$$

$$\frac{d[BrO_2]}{dt} = 0 = 2R_{3a} - R_{3b} \qquad (7\text{-}65)$$

The summation gives

$$R_1 - R_2 + R_{3a} - 2R_4 = 0 \qquad (7\text{-}66)$$

which is an easily solved quadratic in $[HBrO_2]$. We recognize two cases, depending on the relative values of R_1 and R_4.

In the event $R_1 \gg R_4$, the relation gives

$$[HBrO_2]_{ss} = \frac{k_1[BrO_3^-][Br^-]}{k_2[Br^-] - k_{3a}[BrO_3^-]} \qquad (7\text{-}67)$$

Since the condition $R_1 \gg R_4$ will occur at a relatively high value of $[Br^-]_{ss}$, the first denominator term tends to dominate, and the steady-state concentration can be

---

†Clearly, reaction $(5a)$ is not an elementary reaction; it consists of a series of steps with $(5a)$ as a resultant. In γ, $(5a)$ is the rate-limiting process, and it shows a first-order dependence on $[Ce(IV)]$.

approximated

$$[HBrO_2]_{ss,\,small} \simeq \frac{k_1[BrO_3^-]}{k_2} \qquad (7\text{-}68)$$

where the purpose of the notation "small" will be discussed later.

Consider, however, the opposite condition, $R_1 \ll R_4$, in which case Eq. (7-66) yields

$$[HBrO_2]_{ss} = \frac{k_{3a}[BrO_3^-] - k_2[Br^-]}{2k_4} \qquad (7\text{-}69)$$

Under these circumstances, which will prevail at a low value of $[Br^-]_{ss}$, the result is approximated as

$$[HBrO_2]_{ss,\,large} \simeq \frac{k_{3a}[BrO_3^-]}{2k_4} \qquad (7\text{-}70)$$

These results show that the steady-state concentrations of $HBrO_2$ and Br^- are coupled in such a way that, as $[Br^-]_{ss}$ passes through a "critical value," there is a sudden switch from one expression for $[HBrO_2]$ to the other. Based on the denominator of Eq. (7-67) or the numerator of Eq. (7-69), the critical steady-state value is

$$[Br^-]_{ss,\,critical} = \frac{k_{3a}[BrO_3^-]}{k_2} \qquad (7\text{-}71)$$

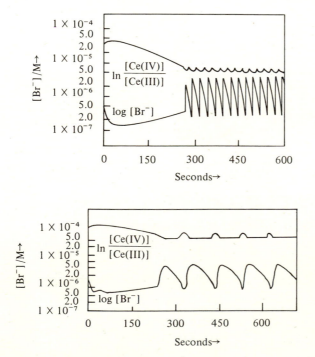

Figure 7-1 Potentiometric traces[3] depicting the oscillation in the ratio $[Ce(IV)]/[Ce(III)]$ and in $[Br^-]$ for two experiments, each with $[Ce(NH_4)_2(NO_3)_6]_0 = 10^{-3}$ M and $[H_2SO_4] = 0.8$ M. The upper experiment has high [MA], with $[CH_2(COOH)_2]_0 = 0.50$ M and $[BrO_3^-]_0 = 0.063$ M; the lower experiment has low bromate content, with $[BrO_3^-]_0 = 0.016$ M and $[CH_2(COOH)_2]_0 = 0.13$ M.

A $[Br^-]_{ss}$ greater than that given by Eq. (7-71) clearly requires Eq. (7-67) to hold, because under such circumstances Eq. (7-69) would give a negative value for $[HBrO_2]$. Likewise, a value of $[Br^-]_{ss}$ which exceeds the critical value causes Eq. (7-69) to apply and renders Eq. (7-67) meaningless.

This switch between the two applicable expressions for $[HBrO_2]$ gives rise to the oscillating phenomenon. Visually, of course, this occurs because the ratio of cerium in oxidation states IV and III is coupled to the other steady-state concentrations. The feedback loop is controlled by $[Br^-]_{ss}$, and a key feature is that $[Br^-]$, too, is oscillating in reverse relation to $[HBrO_2]$ by competition between reactions which form Br^- and those which consume it. Some of these effects are shown in Fig. 7-1, which depicts the oscillation in the ratio $[Ce(IV)]/[Ce(III)]$ and in $[Br^-]$. The figure shows runs under two different conditions, which illustrates how the amplitude and frequency of the oscillations depend upon concentrations.

REFERENCES

1. Boyer, A., M. Niclause, and M. Letort: *J. chim. phys.*, **49**:345 (1952).
2. Brinton, R. K., and D. H. Volman: *J. Chem. Phys.*, **20**:1053 (1952).
3. Benson, S. W.: "The Foundations of Chemical Kinetics," McGraw-Hill Book Company, New York, 1960, p. 383.
4. Ref. 3, pp. 322–325.
5. Noyes, R. M., and R. J. Field: *Accounts Chem. Research*, **10**:273 (1977).
6. Field, R. J., E. Körös, and R. M. Noyes: *J. Am. Chem. Soc.*, **94**:8649 (1952).

PROBLEMS

7-1 *Rate law and activation energy.* Iodine and ethane react at 550 to 600 K according to the equation $I_2 + C_2H_6 = C_2H_5I + HI$. (Ethyl iodide also undergoes slow decomposition to $C_2H_4 + HI$, which is ignored here.) The rate is given by

$$\frac{-d[C_2H_6]}{dt} = k_{exp}[I_2]^{1/2}[C_2H_6]$$

with

$$\log_{10}(k_{exp}/atm^{-1/2} \ s^{-1}) = 12.08 - \frac{43,100 \ cal \ mol^{-1}}{2.3RT}$$

The equilibrium $I_2 \rightleftharpoons 2I\cdot$ has $\Delta H_1^\circ = 33.2$ kcal mol^{-1} and $\Delta S_1^\circ = 24.8$ cal mol^{-1} k^{-1} (at 550 to 600 K, 1 atm standard state).

(*a*) Show that the following mechanism accounts for these results.

(*b*) Express k_2 in Arrhenius form.

$$I_2 + M \underset{-1}{\overset{1}{\rightleftharpoons}} 2I\cdot + M$$

$$I\cdot + C_2H_6 \overset{2}{\longrightarrow} C_2H_5\cdot + HI$$

$$C_2H_5\cdot + I_2 \overset{3}{\longrightarrow} C_2H_5I + I\cdot$$

7-2 *Rate expression.* The net reaction of U(IV) and O_2,

$$2U^{4+} + O_2 + 2H_2O = 2UO_2^{2+} + 4H^+$$

is believed to proceed by the mechanism shown below, balanced only with respect to oxidation-reduction. Derive the expression for $-d[U(IV)]/dt$ by making appropriate steady-state and long-chain approximations.

Initiation:

$$U(IV) + O_2 \xrightarrow{1} HO_2 + U(V)$$

Propagation:

$$U(V) + O_2^- \xrightarrow{2} U(VI) + HO_2$$

$$U(IV) + HO_2 \xrightarrow{3} U(V) + H_2O_2$$

$$U(IV) + H_2O_2 \Longrightarrow U(VI) + H_2O$$

The last reaction is *very fast*; that is, every H_2O_2 produced converts U(IV) to U(VI) directly without forming U(V).

Termination:

$$U(V) + HO_2 \xrightarrow{4} U(VI) + H_2O_2$$

7-3 *Rate expression.* The oxidation of SO_3^{2-} to SO_4^{2-} by molecular oxygen is catalyzed by traces of Cu^{2+} and inhibited by small amounts of alcohols. Derive the rate expression for $-d[SO_3^{2-}]/dt$, by making the usual approximations, assuming the following mechanism:

$$Cu^{2+} + SO_3^{2-} \xrightarrow{1} Cu^+ + SO_3^-\cdot$$

$$SO_3^-\cdot + O_2 \xrightarrow{2} SO_5^-\cdot$$

$$SO_5^-\cdot + SO_3^{2-} \xrightarrow{3} SO_3^-\cdot + SO_5^{2-}$$

$$SO_5^{2-} + SO_3^{2-} \longrightarrow 2SO_4^{2-} \quad \text{(very fast)}$$

$$ROH + SO_5^-\cdot \xrightarrow{4} \text{inert products}$$

7-4 *Rate law and activation energy.* The oxidation of nitrosobenzene (PhNO) by tert-butylhydroperoxide (BuOOH) is catalyzed by di-*tert*-butyl peroxyoxalate (BuOOCOCOOOBu) and by a cobalt(II) chelate complex [Johnson and Gould, *J. Am. Chem. Soc.*, 95:5198 (1973)].

(*a*) In the catalysis by the peroxyoxalate, the chain mechanism is

$$\overset{\displaystyle OO}{\underset{\displaystyle | |}{BuOOCCOOBu}} \xrightarrow{1} 2BuO\cdot + 2CO_2$$

$$BuO\cdot + BuOOH \xrightarrow{3} BuOH + BuO_2\cdot$$

$$BuO_2 + PhNO \xrightarrow{4} BuO\cdot + PhNO_2$$

$$BuO\cdot + PhNO \underset{-2}{\overset{2}{\rightleftharpoons}} Ph(BuO)NO\cdot$$

$$BuO_2\cdot + Ph(BuO)NO\cdot \xrightarrow{5} \text{inactive products}$$

Derive the rate expression. Assume the steady-state approximation for the radical intermediates, and assume that the chains are long.

(*b*) In the catalysis of the same reaction by the cobalt(II) compound the rate law is

$$\frac{-d[\text{PhNO}]}{dt} = k[\text{Co}^{\text{II}}]^{1/2}[\text{BuOOH}]$$

Propose a mechanism for this reaction, and relate k to the rate constants of your mechanism.

(*c*) Designate the composite rate constant from part *a* of this problem as k_a. From the slope of $\ln k_a$ versus $1/T$, an apparent activation energy E_a was computed. What is the algebraic relation between that E_a and the individual activation energies, $E_1, E_2, E_{-2}, E_3, E_4$, and E_5?

7-5 *Termination step.* Add a termination step to the mechanism given below for the photochemical addition of Cl_2 to an olefinic double bond which follows the rate law

$$\frac{-d[\text{olefin}]}{dt} = kI^{1/2}[\text{olefin}]^{1/2}[Cl_2]$$

where I is the intensity of the radiation leading to photodissociation of Cl_2.

Initiation: $$Cl_2 \xrightarrow{h\nu} 2Cl\cdot$$

Propagation: $$Cl\cdot + R_2C{=}CR_2 \xrightarrow{2} R_2C(Cl){-}CR_2\cdot$$

$$R_2C(Cl){-}CR_2\cdot + Cl_2 \xrightarrow{3} R_2C(Cl)CR_2(Cl) + Cl\cdot$$

7-6 *Termination step.* Add a consistent termination step to the mechanism shown for the pyrolysis of ethane at 800 to 1000 K, which occurs according to the stoichiometry $C_2H_6 = C_2H_4 + H_2$ with a $\frac{1}{2}$-order dependence on $[C_2H_6]$. [L. A. Wall and W. J. Moore, *J. Phys. Chem.*, **55**:965 (1951)].

$$C_2H_6 \xrightarrow{1} 2CH_3\cdot$$

$$C_2H_6 + CH_3\cdot \xrightarrow{2} CH_4 + C_2H_5\cdot$$

$$C_2H_5\cdot \xrightarrow{3} C_2H_4 + H\cdot$$

$$H\cdot + C_2H_6 \xrightarrow{4} H_2 + C_2H_5\cdot$$

7-7 *Chain mechanism.* Chlorine catalyzes the decomposition of ozone and follows the rate law

$$\frac{-d[O_3]}{dt} = k[Cl_2]^{1/2}[O_3]^{3/2}$$

Propose a mechanism consistent with this, and relate the experimental rate constant to the individual rate constants of the mechanism.

7-8 *Induced chain reactions.* In acidic solution, oxalic acid reacts very slowly with chlorine, but iron(II) reacts rapidly.

Net reactions:

(1) $$Cl_2 + H_2C_2O_4 = 2H^+ + 2Cl^- + 2CO_2$$

(2) $$Cl_2 + 2Fe^{2+} = 2Fe^{3+} + 2Cl^-$$

The reaction of Fe^{2+} is presumed to have as its first step

(3) $$Cl_2 + Fe^{2+} = Fe^{3+} + Cl^- + Cl\cdot \qquad \text{very rapid}$$

The chlorine atoms generated in this first step are quite reactive; in the presence of oxalic acid they can act as chain carriers for reaction (1) [H. Taube, *J. Am. Chem. Soc.*, **68**:611 (1946)].

Experiment: To a solution of Cl_2 and $H_2C_2O_4$ add a very low concentration of Fe^{2+} at a constant rate R. Since reaction (3) is so fast, the rate of generation of $Cl\cdot$ is equal to R. The chain-carrying steps are presumed to be:

$$HC_2O_4^- + Cl\cdot \xrightarrow{4} C_2O_4^- \cdot + H^+ + Cl^-$$

$$C_2O_4^- \cdot + Cl_2 \xrightarrow{5} 2CO_2 + Cl^- + Cl\cdot$$

The chain-breaking step may be either of the following:

$$2Cl\cdot \xrightarrow{6} Cl_2 \quad \text{or} \quad Cl\cdot + C_2O_4^- \cdot \xrightarrow{7} Cl^- + 2CO_2$$

Assuming chain breaking occurs exclusively by step 6, derive the steady-state rate equation for $-d[H_2C_2O_4]/dt$. Make the steady-state assumption for $Cl\cdot$ and $C_2O_4^- \cdot$, and assume long chains (that is, $[Fe^{2+}]$ added $\ll [Cl_2]$). Answer will involve R. Repeat the derivation for chain termination with step 7.

7-9 *Rate law.* Pratt and Rogers [*J. Chem. Soc.*, *Faraday Trans I*, **12**:2769 (1976)] have studied the exchange reaction $CH_4 + D_2 = CH_3D + HD$ at 957 to 1188 K. They find a rate law

$$\frac{d[CH_3D]}{dt} = k[CH_4][D_2]\left(\frac{1}{[D_2] + C[CH_4]}\right)^{1/2}$$

Show that this rate law is consistent with the chain mechanism shown below, in which two termination steps are of comparable importance. Make the ss approximation and assume the chains are long. Identify k and C for their rate law in terms of the rate constants in the mechanism.

Initiation:	$CH_4 \xrightarrow{1} CH_3 + H$
Fast reaction:	$H + D_2 = HD + D$
Propagation:	$D + CH_4 \xrightarrow{2} HD + CH_3$
	$CH_3 + D_2 \xrightarrow{3} CH_3D + D$
Termination:	$D + CH_3 \xrightarrow{4} CH_3D$
	$2CH_3 \xrightarrow{5} C_2H_6$

Hint: Since H and D are rapidly interconverted, the steady-state equation for D is $d[D]/dt = k_1[CH_4] - k_2[D][CH_4] + k_3[D_2][CH_3] - k_4[D][CH_3] = 0$.

7-10 *Steady-state concentrations.* Using the notation given in the text, the rate constants for the $H_2 + Br_2$ reaction at 600 K are $A = 5.4 \times 10^{-4}$ $M^{-1/2}$ s^{-1} and $B = 0.12$. The value of k_3 from collision theory (which should be quite adequate for such a simple reaction, Sec. 8.1) is 8×10^{11} M^{-1} s^{-1} at 600 K.

(a) Estimate $[H\cdot]_{ss}$ in an experiment with $[H_2] = [Br_2] = [HBr] = 10^{-2}$ M.

(b) Given, for the equilibrium $H_2 = 2H\cdot$ the value $K = 3.3 \times 10^{-35}$ M at 600 K. Estimate $[H\cdot]_{eq}$ under the same conditions, assuming the equilibrium is attained.

(c) Comment perceptively upon $[H\cdot]_{ss}$ and $[H\cdot]_{eq}$ and the bearing that has upon the mechanism.

EIGHT

THEORIES OF ELEMENTARY REACTION RATES

This book is directed toward readers with an interest in chemical reaction mechanisms. Consequently, the treatment of theoretical development is held to a minimum. This chapter is intended to serve as a terse introduction to a subdiscipline of chemical kinetics which abounds with major detailed works on the subject.[1-5] It summarizes the bases of collision theory for bimolecular reactions, activated-complex theory, and the theory of unimolecular reactions.

8-1 COLLISION THEORY FOR BIMOLECULAR REACTIONS

The kinetic theory of gases provides a means of computing a theoretical value for A, the preexponential term of the Arrhenius equation. We start with the equation[6] for $\bar{c}$, the average molecular speed:

$$\bar{c} = \left(\frac{8RT}{\pi Nm}\right)^{1/2} \tag{8-1}$$

where N represents Avogadro's number and m is the molecular mass (kg/molecule). In a mixture of two gases A and B (comprised of spherical molecules of diameters d_A and d_B), a collision is said to occur if molecules contact at the surface of the spheres. The critical collision distance σ is

$$\sigma = \tfrac{1}{2}(d_A + d_B) \tag{8-2}$$

The volume swept out per unit time by a sphere of radius σ, multiplied by $[A][B]$, gives the number of collisions between unlike molecules:

$$Z_{AB} = \pi\sigma^2 \bar{v}_r [A][B] \tag{8-3}$$

where $\bar{v}_r$, *the mean relative velocity* of the two molecules, is somewhat greater than the average velocity $\bar{c}$ of a single molecule. The value of $\bar{v}_r$ is

$$\bar{v}_r = \left(\frac{8RT}{\pi N\mu}\right)^{1/2} \tag{8-4}$$

where μ is the reduced molecular mass and $N\mu$ is the reduced molar mass[†]:

$$\mu = \frac{m_A m_B}{m_A + m_B} \tag{8-5}$$

$$N\mu = \frac{M_A M_B}{M_A + M_B} \tag{8-6}$$

With these substitutions, the expression for the collision frequency becomes[§]

$$Z_{AB} = \left(\frac{8\pi RT}{N\mu}\right)^{1/2} \sigma^2 [A][B] = Z'_{AB}[A][B] \tag{8-7}$$

If this model is correct, the quantity Z'_{AB} should be equal to the preexponential factor in the Arrhenius relation. The idea can be tested by a calculation for the gas-phase reaction of NO and $ClNO_2$, Eq. (6-8). With collision distance σ estimated at 0.35 nm, the value of Z'_{AB} at 298 K is

$$Z'_{AB} = \frac{Z_{AB}}{[A][B]} = \left[\frac{8\pi \times 8.314 \; m^2 \; kg \; s^{-2} \; mol^{-1} \; K^{-1} \times 298 \; K}{(0.030 \times 0.0815)/(0.030 + 0.0815) \; kg \; mol^{-1}}\right]^{1/2}$$

$$\times \frac{(3.5 \times 10^{-10} \; m)^2}{molecule} \tag{8-8}$$

$$Z'_{AB} = 2.1 \times 10^{-16} \; m^3 \; molecule^{-1} \; s^{-1} \tag{8-9}$$

Converting to the units normally used,

$$Z'_{AB} = 2.1 \times 10^{-16} \frac{m^3}{molecule \; s} \times \left(\frac{10^2 \; cm}{m}\right)^3 \times 6 \times 10^{23} \frac{molecule}{mol} \tag{8-10}$$

$$Z'_{AB} = 1.2 \times 10^{14} \; cm^3 \; mol^{-1} \; s^{-1} \tag{8-11}$$

This value is some 200 times larger than the experimental value of A from Eq. (6-9), $6.5 \times 10^{11} \; cm^3 \; mol^{-1} \; s^{-1}$.

[†]The ordinary molecular weights (molar masses) are more convenient for calculational purposes than are molecular masses. All the quantities M, $N\mu$, and Nm are masses per mole (and thus, in SI units, have dimensions (kg mol^{-1}), and the molecular quantities m and μ have units kg molecule^{-1}.

[§]Equation (8-7) needs to be modified for like molecules to avoid a double counting of collisions:

$$Z_{AA} = \frac{1}{2}\left(\frac{8\pi RT}{N\mu}\right)^{1/2} \sigma^2 [A]^2 \tag{8-7a}$$

Before we enquire into the problem, we should note that the agreement between the collision theory values of Z'_{AB} and the experimental values of A are better the smaller and simpler the reactants. Good agreement is usually found for two atoms, and many bimolecular reactions between an atom and a diatomic molecule are also adequately represented.

8-2 STERIC FACTORS AND ENERGY TERMS IN COLLISION THEORY

The larger and more complex the reactants are, the less likely it is that a given collision will occur with an orientation that can lead to reaction. This notion leads to the modification

$$A = pZ'_{AB} \tag{8-12}$$

where p is a *steric factor* to account for the ineffectiveness of many collisions owing to orientation.

Only the collisions whose partners bring sufficient kinetic energy to exceed the critical amount needed to overcome the activation energy are effective, and the total number of collisions in (8-7) is reduced, generally by large amounts, by that energy requirement. Therefore, the expression from collision theory must be multiplied by the factor $\exp(-E^*/RT)$ to yield the actual rate constant. The complete equation is thus

$$\text{Rate} = k[A][B] = pZ_{AB} \exp\left(-\frac{E^*}{RT}\right) \tag{8-13}$$

where the steric factor p $(0 < p < 1)$ reflects the orientational requirement and the exponential reflects the critical energy required.

Although Eq. (8-13) enjoys qualitative success, it is not entirely satisfactory as a model in that p is not given theoretically. Attempts to relate p to molecular structure variations in an ordered series of reactions do not always lead to consistent results. Moreover, there are occasional reactions which would require $p > 1$ to accommodate the experimental findings:

$$CH_3 \cdot + Cl \cdot \longrightarrow CH_3Cl \tag{8-14}$$

$$A = 4.0 \times 10^{14} \text{ cm}^3 \text{ mol}^{-1} \text{ s}^{-1}$$

$$Z'_{AB} = 2.5 \times 10^{13} \text{ cm}^3 \text{ mol}^{-1} \text{ s}^{-1} \ (\Rightarrow p = 16!)$$

One further point should be noted: the hard-sphere model is obviously naive. Real molecules undergo attractions and repulsions and may not approximate spheres in shape. Sufficiently attractive interactions could, for example, account for the occurrence of certain bimolecular reactions faster than the hard-sphere collision rate. Modifications to account for such effects are usually at the expense of the introduction of additional adjustable parameters.

8-3 ACTIVATED COMPLEX THEORY

A wide variety of chemical reactions have been interpreted in terms of activated complex theory (or absolute rate theory[7]). The most extensive application has been in the area of elementary reactions in the gas phase, especially those involving simple molecules whose structural parameters are known. That permits estimation of the requisite partition functions, which, in turn, are used in the calculation of the pseudo-equilibrium constant between reactants and activated complex. With suitable adjustment, the theory has been applied to solution reactions, and one area of notable success is its application by Marcus[8] to rates of electron transfer.

The basic relation, of which an elementary derivation will be given shortly, is the Eyring equation [Eq. (8-15)]. It is used as a parametric relation by which the rate constant

$$k = \frac{RT}{Nh} \exp\left(\frac{\Delta S^{\ddagger}}{R}\right) \exp\left(-\frac{\Delta H^{\ddagger}}{RT}\right) \tag{8-15}$$

for a reaction determined at several temperatures can be correlated by two adjustable parameters $\Delta H^{\ddagger}$ and $\Delta S^{\ddagger}$.

Under the assumption that the theoretical model is valid, one can hope to understand, or rationalize, the values of $\Delta H^{\ddagger}$ and $\Delta S^{\ddagger}$ for a particular reaction or reaction series. The nature and molecular mechanism of the chemical change must be taken into consideration.

The basic premise of the theory is that the reactants exist in a quasi-equilibrium[†] with the activated complex and have a characteristic frequency of passage over the potential-energy maximum or transition state. The scheme thus envisaged is shown in Eqs. (8-16) and (8-17) for a bimolecular reaction of molecules A and B

$$A + B \overset{K}{\rlap{=}{=}} (AB)^{\ddagger} \tag{8-16}$$

$$(AB)^{\ddagger} \overset{\nu}{\longrightarrow} C + D \tag{8-17}$$

The equilibrium step, formulated as if it were a conventional equilibrium, is characterized by an equilibrium constant K and associated thermodynamic functions.

The next section focuses on the formulation which leads to the Eyring equation and on the evaluation of the equilibrium constant in terms of partition functions based on concepts from statistical thermodynamics.

8-4 DERIVATION OF THE ACTIVATED COMPLEX THEORY

An equilibrium[†] constant K can be formulated[9] in terms of the partition functions of reactants and products,

---

[†]The equilibrium is between the reactants and the activated complexes which are crossing the potential-energy surface in the forward direction, i.e., going toward products.

$$K = \frac{[(AB)^{\ddagger}]}{[A][B]} = \frac{Q^{\ddagger}/N}{(Q_A/N)(Q_B/N)} \exp\left(-\frac{\Delta E_0^{\circ}}{RT}\right) \tag{8-18}$$

where Q_i is the partition function of species i, N is Avogadro's number, and ΔE_0° is the change in internal energy at 0 K. The partition function for a molecule containing n atoms can be factored into the product of partition functions for each degree of translational, rotational, and vibrational freedom,

$$Q = q_{tr}{}^3 q_r{}^2 q_v{}^{3n-5} \quad \text{linear molecule} \tag{8-19a}$$

$$Q = q_{tr}{}^3 q_r{}^3 q_v{}^{3n-6} \quad \text{nonlinear molecule} \tag{8-19b}$$

The activated complex is treated like any normal molecule *except* that its vibration along the reaction coordinate has virtually no restoring force. (That vibration completes the reaction, and we designate its particular function as q_v^*.) The general form of the vibrational partition functions summarized in Table 8-1 reduces in the limit $\nu \to 0$ to

$$q_v^* = \frac{1}{1 - \exp(-h\nu/k_B T)} \tag{8-20}$$

$$q_v^* \cong \frac{1}{1 - [1 - h\nu/k_B T + \cdots]} \cong \frac{k_B T}{h\nu} \tag{8-21}$$

Table 8-1 Molecular partition functions

Motion	No. degrees of freedom	Partition function[a]	Typical value[b]	Avg. temp. exponent
Translation	3	$q_{tr}{}^3 = \dfrac{(2\pi m k_B T)^{3/2} V}{h^3}$	10^{30}–10^{33} V[b]	$T^{3/2}$
Rotation (linear molecule)	2	$q_r{}^2 = \dfrac{8\pi^2 I k_B T}{\sigma h^2}$	10–10^2	T^1
Rotation (nonlinear molecule)	3	$q_r{}^3 = \dfrac{8\pi^2 (8\pi^3 ABC)^{1/2} (k_B T)^{3/2}}{\sigma h^3}$	10^2–10^3	$T^{3/2}$
Vibration	1	$q_v = \left[1 - \exp\left(-\dfrac{h\nu}{k_B T}\right)\right]^{-1}$	1–10	$\sim T^0$

[a] $k_B = R/N = $ Boltzmann's constant $= 1.38 \times 10^{-23}$ J K^{-1}; $\sigma = $ rotational symmetry number; $I = $ moment of inertia of a linear molecule $= \frac{1}{2}\mu d^2$; A, B, C $= $ three principal moments of inertia of a nonlinear molecule.

[b] Use V in m^3 for $q_{tr}{}^3$ to be consistent with other numerical values in SI units. Conversion to another standard state requires multiplication by an appropriate conversion factor as illustrated in Eq. (8-39).

The partition function for the activated complex can then be written

$$Q^{\ddagger} = q_v^* Q^* \cong \frac{k_B T}{h\nu} Q^* = \frac{RT}{Nh\nu} Q^* \tag{8-22}$$

where Q^* is the partition function for $(AB)^{\ddagger}$ with the partition function for the reaction coordinate removed. (Replacement of Boltzmann's constant by $k_B = R/N$ avoids possible confusion with a rate constant.) Substitution into Eq. (8-18) gives

$$K = \frac{RT}{Nh\nu} \frac{Q^*/N}{(Q_A/N)(Q_B/N)} \exp\left(-\frac{\Delta E_0^{\circ}}{RT}\right) \tag{8-23}$$

which gives for the rate constant[†]

$$k = \nu K = \frac{RT}{Nh} \left[\frac{Q^*/N}{(Q_A/N)(Q_B/N)} \exp\left(-\frac{\Delta E_0^{\circ}}{RT}\right)\right] \tag{8-24}$$

The quantity in brackets on the right-hand side of Eq. (8-24) is usually written $K^{\ddagger}$, which gives the following relations,[§]

$$k = \frac{RT}{Nh} K^{\ddagger} = \frac{RT}{Nh} \exp\left(-\frac{\Delta G^{\ddagger}}{RT}\right) \tag{8-25}$$

$$k = \frac{RT}{Nh} \exp\frac{\Delta S^{\ddagger}}{R} \exp\left(-\frac{\Delta H^{\ddagger}}{RT}\right) \tag{8-26}$$

8-5 APPLICATION OF ACTIVATED COMPLEX THEORY TO BIMOLECULAR REACTIONS

Consider bimolecular reactions of increasing complexity, starting with a reaction between two atoms A and B. The activated complex, a diatomic molecule, has a special partition function Q^* given by

$$Q^* = q_{tr}^3 q_r^2 \tag{8-27}$$

[†]The rate constant will be in units $m^3 \ mol^{-1} \ s^{-1}$ if V in m^3 is used for q_{tr}^3. If N were omitted, as is sometimes done, k would be units of molecules rather than moles. A similar derivation starting with a unimolecular reaction in place of Eq. (8-16) gives

$$k = \frac{RT}{Nh} \frac{Q^*/N}{Q_A/N} \exp\left(-\frac{\Delta E_0^{\circ}}{RT}\right) \tag{8-24a}$$

$$k = \frac{RT}{Nh} \frac{Q^*}{Q_A} \exp\left(-\frac{\Delta E_0^{\circ}}{RT}\right) \tag{8-24b}$$

[§]It is tempting, considering these relations for the rate constant

$$k = \frac{RT}{Nh} K^{\ddagger} = \nu K$$

to identify ν as RT/Nh and K as $K^{\ddagger}$. That is *not correct*, however, and $K = K^{\ddagger}/q_v^*$, $\nu = RT/Nhq_v^*$. For a molecule with a very loose vibration the approximation $q_v^* \sim 1$ is really quite poor.

because the single vibrational mode is the reaction coordinate separately accounted for by the factor RT/Nh. Substitution of the appropriate quantities yields

$$Q^* = \frac{[2\pi(m_A + m_B)RT/N]^{3/2}}{h^3} \cdot \frac{8\pi^2(m_A m_B/m_A + m_B)d_{AB}^2 RT}{Nh^2} \qquad (8\text{-}28)$$

Substitution into Eq. (8-24) followed by combination of like terms and simplification yields

$$k = \frac{RT}{Nh}\left[\frac{(2\pi RT/N)^{3/2}(8\pi^2 RT/N)}{(2\pi RT/N)^{3/2}(2\pi RT/N)^{3/2}}\right]\left[\frac{m_A m_B(m_A + m_B)^{1/2}}{m_A^{3/2} m_B^{3/2}}\right]$$

$$\times [d_{AB}^2 h] \times \exp\left(-\frac{\Delta E_0^\circ}{RT}\right) \qquad (8\text{-}29)$$

$$k = \frac{RT}{Nh}\left(\frac{8\pi N}{RT}\right)^{1/2}\frac{d_{AB}^2 Nh}{\mu^{1/2}}\exp\left(-\frac{\Delta E_0^\circ}{RT}\right)$$

$$k = \left(\frac{8\pi RT}{N\mu}\right)^{1/2} d_{AB}^2 N \exp\left(-\frac{\Delta E_0^\circ}{RT}\right) \qquad (8\text{-}30)$$

The final expression is identical with that obtained from application of collision theory,[†] Eq. (8-7), where the critical collision distance σ is represented as the distance d_{AB} in the activated complex. The exact agreement is highly encouraging; for we have seen previously that the collision theory model is quite successful in this instance.

For application to the bimolecular reactions, the drastic simplification will be made that all molecules have the same value for q_{tr}, q_r, and q_v; the typical values are those shown in Table 8-1. Actually, the ultimate point of interest is to use activated complex theory to provide a theoretical value for the steric factor p of collision theory. Thus we shall be computing the ratio of two rate constants: the constant for the reaction of interest and a "standard," the value for a reaction of two atoms with $p = 1$. The activation energy terms are presumed to cancel, or in any event they are ignored when the object is to compute the steric factor which enters the preexponential term only.

On such a basis the rate constant for the reaction of two atoms is, from (8-24) and (8-27),

$$k_{2\,\text{atoms}} = \frac{RT}{Nh}\frac{q_{tr}^3 q_r^2/N}{(q_{tr}^3/N)(q_{tr}^3/N)}\exp\left(-\frac{\Delta E_0^\circ}{RT}\right) \qquad (8\text{-}31)$$

$$k_{2\,\text{atoms}} = \frac{RT}{h}\frac{q_r^2}{q_{tr}^3}\exp\left(-\frac{\Delta E_0^\circ}{RT}\right) \qquad (8\text{-}32)$$

[†]Except that (8-30) contains Avogadro's number, which places the rate constant on a per mole basis in contrast to the per molecule basis of (8-7).

Considering first the reaction of an atom with a linear molecule B, having n_B atoms, to form a linear activated complex

$$p = \frac{k}{k_{2\ atoms}} = \frac{\dfrac{RT}{Nh}\ \dfrac{(q_{tr}^3 q_r^2 q_v^{3(n_b+1)-5-1}/N)}{(q_{tr}^3/N)(q_{tr}^3 q_r^2 q_v^{3n_b-5}/N)}}{(RT/h)(q_r^2/q_{tr}^3)} \tag{8-33}$$

$$p = \frac{(RT/h)(q_v^2/q_{tr}^3)}{(RT/h)(q_r^2/q_{tr}^3)} = \frac{q_v^2}{q_r^2} \cong 10^{-1} \text{ to } 10^{-2} \tag{8-34}$$

In a similar manner, the following results can be derived:

Reaction type	Steric factor p	
	Expression	Approx. value[a]
1. $A + B—C \longrightarrow (A—B—C)^{\ddagger}$	q_v^2/q_r^2	10^{-2}
2. $A + B—C \longrightarrow \left(\!\!\begin{smallmatrix} B \\ A \diagdown\ \diagup C \end{smallmatrix}\!\!\right)^{\ddagger}$	q_v/q_r	10^{-1}
3. $A—B + C—D \longrightarrow [A—B—C—D]^{\ddagger}$	q_v^4/q_r^4	10^{-4}
4. $A—B + C\overset{D}{\diagup}\diagdown E \longrightarrow \left[A—B—C\overset{D}{\diagup}\diagdown E\right]^{\ddagger}$	q_v^4/q_r^4	10^{-4}
5. $A\overset{B}{\diagup}\diagdown C + D\overset{E}{\diagup}\diagdown F \longrightarrow \left[A\overset{B}{\diagup}\diagdown C\overset{D}{\diagup}\diagdown E\overset{F}{\diagdown}\right]^{\ddagger}$	q_v^5/q_r^5	10^{-5}

[a]Using $q_v \sim 1$ and $q_r \sim 10$.

8-6 CALCULATIONS BASED ON ACTIVATED COMPLEX THEORY

Following Skinner,[10] we consider a calculation of the preexponential factor (that is, $\Delta S^{\ddagger}$) for the elementary bimolecular reaction of Eq. (8-35), which has been studied experimentally at 20 to 60°C.[11]

$$Br + Cl_2 \longrightarrow BrCl + Cl \tag{8-35}$$

$$k = \frac{RT}{Nh} \exp\left(-\frac{66 \pm 5 \text{ J mol}^{-1}\text{ k}^{-1}}{R}\right) \exp\left(-\frac{27.1 \pm 1.7 \text{ kJ mol}^{-1}}{RT}\right) \tag{8-36}$$

Comparison of Eqs. (8-24) and (8-26) leads to the expression

$$\Delta S^{\ddagger} = R \ln \frac{Q^*/N}{(Q_{Br}/N)(Q_{Cl_2}/N)} \tag{8-37}$$

and based on this relation the values of $\Delta S^{\ddagger}$ will be calculated at 300 K.

Consider first the partition function for atomic bromine. The only partition function for Br is that for translation; the value for a standard state of 1 mol dm^{-3} is calculated as follows:

$$\frac{Q_{Br}}{N} = \frac{q_{tr}{}^3}{N} = \frac{(2\pi m_{Br} kT)^{3/2}}{Nh^3} \tag{8-38}$$

$$\frac{Q_{Br}}{N} = \frac{(2\pi \times 1.33 \times 10^{-25} \text{ kg} \times 1.38 \times 10^{-23} \text{ J K}^{-1} \times 300 \text{ K})^{3/2}}{6.02 \times 10^{23} \text{ mol}^{-1} \times (6.62 \times 10^{-34} \text{ J s})^3} \times \frac{10^{-3} \text{ m}^3}{\text{dm}^3} \tag{8-39}$$

$$\frac{Q_{Br}}{N} = 1.17 \times 10^6 \quad \text{(at 1 mol dm}^{-3}) \tag{8-40}$$

The partition function for Cl_2 consists of contributions from rotation and vibration as well as translation:

$$\frac{Q_{Cl_2}}{N} = \frac{q_{tr}{}^3}{N} q_r{}^2 q_v \tag{8-41}$$

Based on Eq. (8-38), $q_{tr}{}^3/N = 9.8 \times 10^5$ at 1 mol dm^{-3}. The value of $q_r{}^2$, based on an internuclear distance of 1.99×10^{-10} m in Cl_2 is given by

$$q_r{}^2 = \frac{8\pi^2 \times \frac{1}{2} \times 5.9 \times 10^{-26} \text{ kg} \times (1.99 \times 10^{-10} \text{ m})^2 \times 1.38 \times 10^{-23} \text{ J K}^{-1} \times 300 \text{ K}}{2 \times (6.62 \times 10^{-34} \text{ J s})^2} \tag{8-42}$$

$$q_r{}^2 = 436 \tag{8-43}$$

The vibrational frequency of Cl_2 is 557 cm^{-1} (or 557 cm^{-1} $\times 10^2$ cm m^{-1} $\times$ 2.997×10^8 m s^{-1} = 1.67×10^{13} s^{-1}), giving the vibrational partition function as

$$q_v = \frac{1}{1 - \exp\left(\dfrac{-6.62 \times 10^{-34} \text{ J s} \times 1.67 \times 10^{13} \text{ s}^{-1}}{1.38 \times 10^{-23} \text{ J K}^{-1} \times 300 \text{ K}}\right)} \tag{8-44}$$

$$q_v = 1.07 \tag{8-45}$$

The total partition function for Cl_2 is thus, according to Eq. (8-41), 4.6×10^8 at 1 mol dm^{-3}.

The calculation for $BrCl_2$ involves some estimates. It will perhaps be linear, analogously to several $X_3{}^-$ ions, but with one less bonding electron may have longer bonds and looser vibrations. The value of $(q_{tr}{}^3/N)$ is 3.0×10^6 at 1 mol dm^{-3}. If one takes the bond distances as some 25 percent longer than those of Cl_2 and ClBr $(1.99 \times 10^{-10}$ and 2.14×10^{-10} m, respectively), $q_r{}^2 = 8.4 \times 10^3$. Skinner[10] uses tabulated[12] vibrational frequencies for $X_3{}^-$ molecules to provide a rough estimate of ν_1 (symmetric stretching) and ν_2 (the doubly degenerate bonding mode) as 190 and 110 cm^{-1}, respectively. (Of course, ν_3 for antisymmetric stretching is ignored, because it corresponds to the reaction coordinate.) These values give $q_v{}^3 \simeq 10$. Hence $Q_{Br^*Cl_2}/N = 2.5 \times 10^{11}$ at 1 mol dm^{-3}.[†]

[†]For the triplet species Br· and $(BrCl_2)^{\ddagger}$, the contributions from an electronic partition function cancel one another and have thus been omitted.

The results, when inserted into Eq. (8-37), give the value

$$\Delta S^{\ddagger} = 8.314 \text{ J mol}^{-1} \text{ K}^{-1} \ln \frac{2.5 \times 10^{11}}{(1.17 \times 10^{6})(4.6 \times 10^{8})} \tag{8-46}$$

$$\Delta S^{\ddagger} = -64 \text{ J mol}^{-1} \text{ K}^{-1} \tag{8-47}$$

which is in excellent agreement with the experimental value $\Delta S^{\ddagger}$ - 65.9 J mol^{-1} K^{-1}, especially considering the uncertainty in the latter is ± 5.0 J mol^{-1} K^{-1}. The reader can recognize that this was an especially favorable situation, however, and that in reactions of less well characterized or more complex molecules considerably less precision can be hoped for. Also, such calculations are largely limited to the gas phase, because partition functions for species in solution generally are not known.

8-7 KINETICS OF UNIMOLECULAR GAS-PHASE REACTIONS

All unimolecular gas-phase reactions exhibit an interesting kinetic phenomenon: the apparent rate constant, although unvarying at "high pressure," below a certain pressure decreases and continues to decrease in proportion to pressure. Thus the reaction follows second-order kinetics in the "fall-off region." As an illustration consider the results reported for the kinetics of the reaction in Eq. (8-48), the isomerization of methyl isocyanide to acetonitrile.

$$CH_3NC \longrightarrow CH_3CN \tag{8-48}$$

The apparent first-order rate constant, k_{obs}, can be defined as $-d \ln [CH_3NC]/dt$. Values were determined at many different pressures at a constant temperature.[13] A representative sample of an extensive set of determinations at one temperature is given in Table 8-2. It is seen that the rate constant does, indeed, approach a constant value

Table 8-2 Pressure dependence of the apparent first-order constant for reaction (8-48)[a]

P/torr	$10^6 \, k_{obs}$/s^{-1}	P/torr	$10^6 \, k_{obs}$/s^{-1}
0.161	1.01	35.7	34.0
0.297	1.36	72.3	45.0
0.648	2.83	101.5	52.0
1.03	4.08	155.2	55.6
2.03	5.91	208	60.2
2.64	7.38	570	69.8
5.00	11.7	1143	73.0
7.25	14.2	2248	73.7
10.0	18.0	5010	74.1
15.7	21.6	7106	74.9
24.8	30.0		

[a]Data from Ref. 13. Only a representative sample of the published data is given here; the values refer to $T = 199.4°$C.

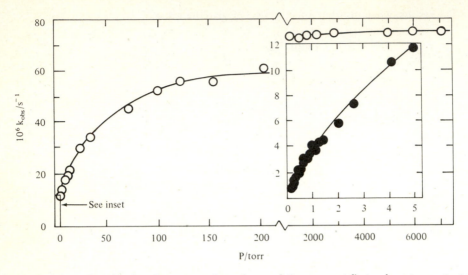

Figure 8-1 A plot showing the pressure dependence of the apparent first-order rate constant for Eq. (8-48); data from Ref. 13 and Table 8-2.

at high pressures but falls off markedly at lower pressures. The data are also displayed graphically in Fig. 8-1.

Coupled with this phenomenon is the following conceptual problem: How does the reactant molecule acquire sufficient energy for reaction? If the activation were to be by collision, which is the only process one readily envisages, second-order kinetics would be expected.

The mechanism proposed by Lindemann in 1922, although not completely correct in the original form, has been the basis of modern theories of unimolecular reactions.

The Lindemann mechanism is

$$A + A \underset{-1}{\overset{1}{\rightleftharpoons}} A^* + A \tag{8-49}$$

$$A^* \overset{2}{\longrightarrow} B + C \tag{8-50}$$

The steady-state approximation for $[A^*]$ gives

$$\frac{-d[A]}{dt} = \frac{k_1 k_2 [A]^2}{k_2 + k_{-1}[A]} \tag{8-51}$$

$$k_{app} = \frac{k_1 k_2 / k_{-1}}{1 + k_2 / k_{-1}[A]} = \frac{k_\infty}{1 + k_2 / k_{-1}[A]} \tag{8-52}$$

where the high-pressure rate constant, referring to the pressure region in which exact first-order kinetics is followed, is $k_\infty = k_1 k_2 / k_{-1}$.

The resulting expression readily accounts for the quantitative features cited, but it suffers from two deficiencies: (1) The pressure region, or $[A]$, in which the reaction is expected to enter the fall-off region is predicted to be many orders of magnitude larger than observed, the latter being typically 10 to 10^2 torr. (2) The linear plot of $1/k_{app}$ versus $1/[A]$ predicted by Eq. (8-52) is often curved. For example, the pressure at which the rate constant has fallen to one-half its limiting value, according to the Lindemann equation, is $P_{1/2} = RTk_\infty/k_1$ (see Prob. 8-8). A value can be calculated for reaction (8-48) if k_1 is estimated at this temperature (199.4°C) by collision theory and k_∞ from the experimental data given in Table 8-2 is 7.5×10^{-5} s^{-1}. The value of k_∞ at other temperatures leads to an activation energy 160 kJ mol^{-1}. This value is to be identified as $E_1 - E_{-1} + E_2$, but the last two terms will contribute little because the activated molecule A* needs little further activation. The value of Z_{AA} from the collision theory expression of Eq. (8-7a) with $\sigma = 4.5 \times 10^{-10}$ m is 1.3×10^{11} dm^3 mol^{-1} s^{-1}. The expected pressure $P_{1/2}$ is thus

$$P_{1/2} = \frac{RTk_\infty}{k_1} \tag{8-53}$$

$$P_{1/2} = \frac{8.314 \text{ J mol}^{-1} \text{ K}^{-1} \times 473 \text{ K} \times 7.5 \times 10^{-5} \text{ s}^{-1}}{1.3 \times 10^8 \text{ m}^3 \text{ mol}^{-1} \text{ s}^{-1} \exp\left(-\dfrac{160{,}000 \text{ J mol}^{-1}}{8.314 \text{ J mol}^{-1} \text{ K}^{-1} \times 473 \text{ K}}\right)} \tag{8-54}$$

$$P_{1/2} = 1.0 \times 10^9 \text{ Pa} \cong 10^4 \text{ atm} \tag{8-55}$$

which can be seen to be terribly far from the experimental value. These considerations illustrate that the Lindemann formulation, although resulting in the correct qualitative description, is inadequate in quantitative detail.

One further line of experimental evidence should be considered; it is that concerning the effects of nonreactive gas molecules other than the reactant itself. Addition of such gases, symbolized as M, gives rise in effects qualitatively resembling those resulting from increasing the concentration of the reactant A. That is to say, the value of k rises from low [M] and attains a constant, limiting value at high [M]. Interestingly enough, the efficiency of each molecule M is different. This phenomen can be considered in terms of the Lindemann mechanism:

$$A + M \underset{k_{-1}^M}{\overset{k_1^M}{\rightleftharpoons}} A^* + M \tag{8-56}$$

$$A^* \xrightarrow{k_2} B + C \tag{8-57}$$

$$k_{obs} = \frac{k_1^M k_2 [M]}{k_{-1}^M [M] + k_2} \tag{8-58}$$

One way to express the results is to compare the effectiveness of various molecules M at low pressures. In this limit $k_{obs} \simeq k_1^M [M]$, and one can cite values of k_1^M/k_1^A. The isomerization of methyl isocyanide, Eq. (8-48), has been studied in the presence of 102 different added gases[14]. Table 8-3 gives the value of $k_1^M/k_1^{CH_3NC}$ for selected gases. Perhaps more immediately useful is the ratio corrected to a "per collision" basis using the reduced masses and molecular radii. Values of $k_1^M(\mu_{AM})^{1/2}\sigma_{AM}/k_1^A(\mu_{AA})^{1/2}\sigma_{AA}$ are also shown. It is clear that the least effective molecules are monoatomic gases and the most effective are large and complicated polyatomics. Similar results have been

Table 8-3 Relative efficiencies of various added gases M on the rate of isomerization of methyl isocyanide, A^a

M	$\dfrac{k_1{}^M}{k_1{}^A}$	$\dfrac{k_1{}^M}{k_1{}^A}\left(\dfrac{\mu_{AM}}{\mu_{AA}}\right)^{1/2}\dfrac{\sigma_{AA}}{\sigma_{AM}}$
Xe	0.119	0.232
He	0.171	0.24
H_2	0.28	0.24
Ne	0.120	0.276
Ar	0.136	0.279
N_2	0.121	0.38
CO	0.25	0.46
CO_2	0.32	0.56
CH_4	0.44	0.61
CF_3H	0.49	0.71
C_3H_8	0.62	0.79
C_2H_5CN	0.88	0.88
CD_3CN	0.92	0.91
$n\text{-}C_8H_{18}$	1.05	0.96
$n\text{-}C_5H_{11}CN$	1.17	0.99
CF_3COCH_3	0.92	1.0
CH_3NC	(1.00)	(1.00)
$n\text{-}C_{10}H_{22}$	1.28	1.02

[a]S. C. Chan, B. S. Rabinovitch, L. D. Spicer, T. Fujimoto, Y. N. Lin, and S. P. Pavlou, *J. Phys. Chem.*, 74:3160 (1962).

found for other unimolecular reactions. The physical basis may be considered as follows. The relative values of $k_1{}^M$ are directly proportional to those for collisional deactivation, $k_{-1}{}^M$, since their ratio is simply the Boltzmann equilibrium ratio of A^* to A and is independent of M. An atom or small molecule is relatively ineffective at collisional deactivation of A^*; hence, on the average A^* will have a greater lifetime and therefore a higher probability of decomposing before being deactivated by further collision.

The most successful modification is that formulated by Rice, Ramsperger, and Kassel and later modified by Marcus. The RRKM theory, as it is commonly referred to, replaces the simple k_2 of Eq. (8-50) by an energy-dependent quantity $k_{2(E)}$. The qualitative rationale can be expressed as follows: An energized molecule A^* with an energy greatly in excess of the minimum (E^*) required for reaction is more likely to react unimolecularly than will a molecule A^* whose energy is barely in excess of E^*.

The RRKM expression is

$$k_{2(E)} = k^{\ddagger}\left(\frac{E - E^*}{E}\right)^{s-1} \tag{8-59}$$

where $k^{\ddagger}$ is the rate constant for free passage over the potential barrier and s is the number of modes of vibrational freedom in which excess energy can be stored.

Proper development of the RRKM theory is beyond the scope of this text, and the interested reader is referred to pertinent references.[2,4,8,15]

REFERENCES

1. Glasstone, S., K. J. Laidler, and H. Eyring: "The Theory of Rate Processes," McGraw-Hill Book Company, New York, 1941.
2. Laidler, K. J.: "Theories of Chemical Reaction Rates," McGraw-Hill Book Company, New York, 1969.
3. Levine, R. D., and R. B. Bernstein: "Molecular Reaction Dynamics," Oxford University Press, New York, 1974.
4. Johnston, H. S.: "Gas Phase Reaction Rate Theory," The Ronald Press Company, New York, 1966.
5. Hirschfelder, J. O., and D. Henderson (eds.): "Chemical Dynamics. Papers in Honor of Henry Eyring," Wiley-Interscience, New York, 1971.
6. See, for example, W. J. Moore, "Physical Chemistry," 3d ed., Prentice-Hall, Inc., Englewood Cliffs, N.J., 1962, chap. 7.
7. Eyring, H., J. Walter, and G. E. Kimball: "Quantum Chemistry," John Wiley & Sons, Inc., New York, 1944, chap. 16.
8. Marcus, R. A.: in Lewis, pp. 35–39, and references therein.
9. Moore, W. J.: "Physical Chemistry," 3d ed., Prentice-Hall, Inc., Englewood Cliffs, N.J., 1962, pp. 635–637.
10. Skinner, G. B.: "Introduction to Chemical Kinetics," Academic Press, Inc., New York, 1974, pp. 64–67.
11. Christie, M. L., R. S. Roy, and B. A. Thrush: *Trans. Faraday Soc.*, 55:1139 (1959).
12. Nakamoto, K.: "Infrared Spectra of Inorganic and Co-ordination Compounds," 2d ed., John Wiley & Sons, Inc., New York, 1970.
13. Schneider, F. W., and B. S. Rabinovitch: *J. Am. Chem. Soc.*, 84:4215 (1962).
14. Chan, S. C., B. S. Rabinovitch, L. D. Spicer, T. Fujimoto, Y. N. Lin, and S. P. Pavlou: *J. Phys. Chem.*, 74:3160 (1962).
15. Robinson, P. J., and K. A. Holbrook: "Unimolecular Reactions," Wiley-Interscience, New York, 1972.

PROBLEMS

8-1 *Collision theory.* Consider $O_2(g)$ at 298 K and $p = 1$ atm $= 10^5$ N m^{-2}.

(a) Compute the number of binary collisions per cm^3 per s, assuming a collision diameter of 3.0×10^{-10} m.

(b) Compute the concentration of O_4 molecules if the duration of a collision is 10^{-13} s.

8-2 *Collision theory and activated complex theory.* Refer to the reactions in Prob. 7-9; Pratt and Rogers find that the Arrhenius preexponential factor for the elementary reaction

$$CH_3 + D_2 \xrightarrow{\ 3\ } CH_3D + D$$

is $A_3 = 1.6 \times 10^{12}$ cm^3 mol^{-1} s^{-1} in the vicinity of 1000 K.

(a) Estimate the steric factor by using approximate values of partition functions.

(b) Combine this p and the experimental data to estimate the critical collision distance σ in the activated complex.

8-3 *Activated complex theory.* What is the temperature dependence of the preexponential factor of each of the five reaction types tabulated in Sec. 8-5?

8-4 *Activated complex theory.* Consider a bimolecular reaction between an atom and a nonlinear molecule to form a nonlinear activated complex:

$$A + B \overset{C}{\diagup}\!\!\diagdown_{D} \longrightarrow \left(\overset{A}{\diagdown}\underset{B}{\diagup}\overset{C}{\diagup}\!\!\diagdown_{D} \right)^{\ddagger}$$

(a) What do you expect as the temperature dependence of the preexponential factor?

(b) Estimate A at 300 K in units $cm^3 \ mol^{-1} \ s^{-1}$.

(c) What is the expression in terms of partition functions, and what is the approximate value, of the steric factor?

8-5 *Unimolecular reactions.* Pritchard, Snowden, and Trotman-Dickenson [*Proc. Roy. Soc. London*, A217:563 (1953)] have studied the isomerization of cyclopropane to propylene. Use the data tabulated below for $T = 743$ K to (a) test its agreement with the Lindemann equation and (b) estimate k_∞.

$10^{-5} \ p/N \ m^{-2}$	1.00	0.510	0.276	0.145	0.067	0.034
$10^4 \ k/s^{-1}$	1.11	1.08	1.04	0.96	0.84	0.79

8-6 *Unimolecular reactions.* Schneider and Rabinovitch [*J. Am. Chem. Soc.*, 84:4215 (1962)] have studied the unimolecular gas-phase isomerization of methyl isocyanide to acetonitrile. The value of the high-pressure rate constant, written in collision theory format for 480 to 520 K, is

$$k_\infty/s^{-1} = 1.1 \times 10^{12} \ T^{1/2} \ \exp\left(-\frac{158,600 \ J \ mol^{-1}}{RT}\right)$$

For $T = 504$ K.

(a) Compute k_{-1} from collision theory ($E_{-1} = 0$ for collisional deactivation); take a collision distance $\sigma = 4.5 \times 10^{-10}$ m.

(b) Compute k_2, noting that $Z'_1 = Z'_{-1}$ and that all the activation energy of k_∞ is attributed to k_1 [i.e., $k_1 = k_{-1} \exp(-158,600/RT)$].

(c) From the Lindemann treatment, estimate the $[CH_3NC]$ at which k_{app} will fall to one-half of k_∞.

(d) Actually, $[CH_3NC]_{1/2}$ is *much* lower than in (c), $2.6 \times 10^{-6} \ mol \ cm^{-3}$. What value of k_1 does that give?

(e) If the discrepancy between k_1 values were to be reconciled by Hinshelwood's expression

$$k_1 = k_{-1} \frac{1}{(S-1)!} \left(\frac{E^*}{RT}\right)^{s-1} \exp\left(-\frac{E^*}{RT}\right)$$

(which assumes that the only effective collisional energy is that stored in certain of the vibrational modes, s being the number of effective modes), compute the value of s needed (by approximation to the nearest whole number).

8-7 *Unimolecular reactions, RRKM theory.* Consider the RRKM expression for $k_{2(E)}$. The energy distribution of activated molecules is

$$P(E) = \left(\frac{E}{RT}\right)^{s-1} \frac{\exp(-E/RT)}{(s-1)!RT}$$

The value of k_∞ is proportional to the product $k_{2(E)}P(E)$, one factor of which falls rapidly with E and the other of which rises sharply. Consequently, the expected value of k_∞ goes through a maximum, and the energy at which that occurs is called E_{max}. Express E_{max} in terms of E^*, s, and RT.

8-8 *Unimolecular reactions.* Express $p_{1/2}$ (the pressure at which the rate has fallen to one-half of k_∞) in terms of k_∞ and k_1 in the Lindemann equation. Calculate $p_{1/2}$ for the unimolecular decomposition of ethane, where $k_\infty = 1.3 \times 10^{-5} \ s^{-1}$ (at 900 K), $E_a = 380 \ kJ \ mol^{-1}$. (Assume that $E^* > E_a$, and take $Z'_{AA} \cong 10^{14} \ cm^3 \ mol^{-1} \ s^{-1}$.)

8-9 *Chain reactions; collision theory; activated complex theory.* Walsh and Benson [*J. Am. Chem. Soc.*, 88:4570 (1966)] report that the net reaction $I_2 + CH_2O = 2HI + CO$ follows a rate law in accord with this mechanism:

$$I_2 + M \underset{-1}{\overset{1}{\rightleftarrows}} 2I + M$$

$$I + CH_2O \xrightarrow{2} CHO + HI$$

$$CHO + I_2 \xrightarrow{3} CO + HI + I$$

They give this expression for k_2, evaluated for 180 to 300°C:

$$k_2/cm^3 \ mol^{-1} \ s^{-1} = 8.3 \times 10^{13} \ exp\left(-\frac{72,500 \ J \ mol^{-1}}{RT}\right)$$

(a) Derive the steady-state rate law.

(b) Compare k_2 with e value derived from collision theory ($\sigma = 4.0 \times 10^{-10}$ m).

(c) Estimate the steri factor p, and again compare theory and experiment.

8-10 *Activated complex theory.* If the equation of activated complex theory, Eq. (8-26), is applied to a bimolecular reaction, there appears to be a discrepancy in the units; the left-hand side has dimensions concentration^{-1} time^{-1}, whereas the right-hand side has time^{-1}. Comment.

NINE

REACTIONS IN SOLUTION

Chemical reactions are most often conducted in the liquid phase, and most studies of reaction mechanisms in organic and inorganic systems concern solution reactions. That has been true to the present time despite the potentially greater depth of theoretical understanding of gas-phase processes.

The present chapter considers a number of issues that have to do with reactions in solution. Some are of a theoretical nature concerning the diffusion rates of solutes and applications of activated complex theory to questions of solvent effects and reactivity; others are of a more practical bent dealing with salt effects of inert electrolytes and rate changes at very high pressures.

9-1 THE NATURE OF REACTIONS IN A SOLVENT

Relatively few reactions are amenable to study in both the gas phase and a variety of solvents. Among them is the decomposition of N_2O_5:

$$N_2O_5 = 2NO_2 + \tfrac{1}{2}O_2 \qquad (9\text{-}1)$$

The formation of tetraethylammonium iodide, although not a suitable gas-phase reaction, can be conducted in many polar solvents:

$$Et_3N + EtI = Et_4N^+I^- \qquad (9\text{-}2)$$

Kinetic data for reactions (9-1) and (9-2) are summarized in Table 9-1. On the one hand, the rate of N_2O_5 decomposition can be seen to be the same in the gas phase

Table 9-1 Solvent effects on reaction rates

Reaction (9-1): $N_2O_5 = 2NO_2 + \frac{1}{2}O_2$

Solvent	$10^5\ k_{298}/s^{-1}$	$\log_{10} A$	$E_a/kJ\ mol^{-1}$
Gas phase	3.4	13.6	103
CCl_4	4.1	13.8	107
$CHCl_3$	3.7	13.6	103
$C_2H_2Cl_2$	4.8	13.6	102
C_2HCl_5	4.3	14.0	105
CH_3NO_2	3.1	13.5	102
Br_2	4.3	13.3	103
$N_2O_4(l)$	7.1	14.2	105
$HNO_3(l)$	0.15	14.8	118

Reaction (9-2): $Et_3N + EtI = Et_4N^+I^-$

Solvent	Dielectric constant, ϵ	$10^5\ k_{373}/M^{-1}\ s^{-1}$	$\log_{10} A$	$E_a/kJ\ mol^{-1}$
n-Hexane	1.9	0.5	4.0	67
Toluene	2.40	25.3	4.0	77
Benzene	2.23	39.8	3.3	48
p-Dichlorobenzene	2.86	70	4.5	53
m-Dichlorobenzene	4.9	111	–	–
Fluorobenzene	5.4	116	3.9	48.9
Bromobenzene	–	166	4.6	52.3
o-Dichlorobenzene	9.9	250	5.1	54.4
Acetone	21.4	265	4.4	49.8
Benzonitrile	25.2	1120	5.0	49.8
Nitrobenzene	36.1	1380	4.9	48.5

as in solution and nearly invariant from one solvent to the next. On the other hand, reaction (9-2) is quite sensitive to changes in solvent; it is faster the more polar the solvent. (A quantitative treatment will be given in Sec. 9-4.) The latter reaction, considering its nature, would be likely to have a polar transition state, and it would be expected to be stabilized in solvents of high dielectric constant. These data serve to show that no single simple generalization can be drawn: In some cases there may be reactions which proceed seemingly independently of solvent, but in general that will not be the case.

It has been pointed out[1] that the mean distance between two solutes at 0.02 mol/dm³ concentration is roughly the same as in the gas phase at one atmosphere pressure, the separation being ~10 molecular diameters. Formation of such an aqueous solution, in which the solvent:solute ratio will be ca. 10^3:1, will necessarily reduce the mean free paths of the reactants to the order of one molecular diameter.

Rather than collisions between noninteracting partners, as in the gas phase, solutes in solution tend to undergo multiple collisions within a solvent cage; this event is referred to as an "encounter."

9-2 THE RATES OF DIFFUSION-CONTROLLED REACTIONS

The maximum rate at which two solutes can react is controlled by the relative rate of diffusion or, more precisely, the encounter rate of the solutes. The mathematical model will have the same significance for solution reactions as collision theory has for gaseous reactants, although the mathematical formulation is rather different. A simple approach[2,3] involving Fick's laws of diffusion will be given, although more rigorous formulations are possible.[4]

The quantity of solute B crossing a plane of area A in unit time (1 s) is a flux designated by J or dn_B/dt, and it has the units molecules s^{-1}. Fick's first law states that this flux is inversely proportional to the gradient in concentration with distance and is, of course, negative, since the flow will be in a direction to offset the gradient:

$$J = \frac{dn_B}{dt} = -DA \frac{dc}{dx} \tag{9-3}$$

where the proportionality constant D is called the diffusion constant. A typical solute has $D \approx 1 \times 10^{-9}$ m^2 s^{-1}, and in practice, values do not vary greatly either with solute or with solvent for solvents of ordinary viscosity.

Consider the bimolecular reaction of solutes A and B, assumed to be uniformly distributed throughout the solution. The concentration of B is depleted near the still unreacted A by virtue of the rapid reaction of A and B, which creates a concentration gradient. It is assumed that there is a critical distance r_{AB}, approximately the sum of molecular radii $r_A + r_B$, at which reaction occurs. At distances $r \leqslant r_{AB}$, [B] = 0; beyond this distance [B] = [B]^0 at $r > r_{AB}$, where [B]^0 symbolizes the bulk concentration of B at $r = \infty$.

Provided the bulk concentrations remain constant, the flux J of Eq. (9-3) is also constant. It is assumed that there is a spherical distribution of potential reactants (B) around a particular molecule of A and that the distance r is measured from A. The surface area of a sphere a distance r from A is $4\pi r^2$, and the expression for the flow of B toward A is

$$J = -D_{AB}(4\pi r^2) \frac{d[B]}{dr} \tag{9-4}$$

where D_{AB} represents the diffusion coefficient for relative diffusion of reactive molecules.

Integration of Eq. (9-4) between the indicated limits affords this result:

$$J \int_{r_{AB}}^{\infty} \frac{dr}{r^2} = -4\pi D_{AB} \int_{0}^{[B]^0} d[B] \tag{9-5}$$

$$\frac{J}{r_{AB}} = -4\pi D_{AB}[B]^0 \tag{9-6}$$

The rate of this bimolecular reaction is given by $-d[A]/dt = k_{obs}[A][B]^0$, but that is also equal to the flux of B toward A, $-J$, multiplied by [A], or $-d[A]/dt = -J[A]$.

Substitution of Eq. (9-6) into these relations gives

$$k_{obs} = \frac{-d[A]/dt}{[A][B]^0} = \frac{-J[A]}{[A][B]^0} \tag{9-7}$$

$$k_{obs} = 4\pi r_{AB} D_{AB} \tag{9-8}$$

on a molecular basis, or

$$k = 4\pi N r_{AB} D_{AB} \tag{9-9}$$

on a molar basis. This latter relation is sometimes referred to as the Smoluchowski equation. Substituting typical values, $r_{AB} = 4 \times 10^{-10}$ m and $D_{AB} = 2 \times 10^{-9}$ m^2 s^{-1},

$$k = 4\pi \times 6 \times 10^{23} \frac{\text{molecules}}{\text{mol}} \times 4 \times 10^{-10} \frac{\text{m}}{\text{molecule}} \times 2 \times 10^{-9} \frac{\text{m}^2}{\text{s}} \times \frac{10^3 \text{ dm}^3}{\text{m}^3}$$

$$k = 6 \times 10^9 \text{ dm}^3 \text{ mol}^{-1} \text{ s}^{-1}$$

This value, often taken as ca. 10^{10} M^{-1} s^{-1}, is referred to as the diffusion-controlled rate constant. It is some 30 times lower than the calculated value of the comparable parameter for gas-phase collision rates, Z'_{AB}.

For interacting particles (ions or dipoles) an additional factor must be incorporated into Eq. (9-9) to account for attractive or repulsive forces, which gives

$$k = 4\pi DN \frac{Nz_A z_A}{\epsilon RT} \frac{1}{\exp(z_A z_A N/\epsilon RT\sigma_{AB}) - 1} \tag{9-10a}$$

This equation, for oppositely charged ions, becomes

$$k \simeq 4\pi DN \frac{Nz_A z_B}{\epsilon RT} \tag{9-10b}$$

where the z's are the ionic charges and ϵ is the dielectric constant of the solvent. For water at 298 K the ratio as a function of $z_A z_B$ and σ_{AB} is as shown in Table 9-2.

This calculation places an approximate upper limit, roughly 10^{10} dm^3 mol^{-1} s^{-1}, on the rate constant for a bimolecular reaction in solution. Reactions of the hydrated

Table 9-2 Influence of ionic charge on the diffusion-controlled rates of second-order reactions

Entries are $k_{ions}/k_{molecules}$

$z_A z_B$	$10^{10} \sigma_{AB}/$m		
	2.00	5.00	10.0
+2	0.005	0.17	0.45
+1	0.10	0.45	0.69
−1	3.7	1.9	1.4
−2	7.1	3.0	1.9

proton, a species having an unusually high diffusion constant, can exceed that value. For example, the reaction $H^+(aq) + OH^- \longrightarrow H_2O$ has $k = 1.4 \times 10^{11}$ dm^3 mol^{-1} s^{-1}, probably the highest value known. Because of the limitation, it is sometimes possible to make a distinction between alternatives which would normally be impossible from kinetics. An illustration of this consideration in distinguishing between alternative mechanisms was given in Sec. 5-3.

9-3   APPLICATIONS OF ACTIVATED COMPLEX THEORY

The basic formulation developed in Chap. 8 for gas-phase reactions is written

$$k_{gas} = \frac{RT}{Nh} K^{\ddagger}$$

(9-11)

Effects due to, say, changes in solvent or solvent composition or ionic strength arise from changes in activity coefficients, a situation which does not assume major importance in gases.

The general method for handling such cases will be given here, and applications will be given in subsequent sections. The value of $K^{\ddagger}$ can be written in terms of activities a and related to that containing concentrations by appropriate activity coefficients. It is

$$K^{\ddagger}_{actual} = \frac{a^{\ddagger}}{a_A a_B} = \frac{[(AB)^{\ddagger}]}{[A][B]} \times \frac{\gamma^{\ddagger}}{\gamma_A \gamma_B}$$

Substitution into (9-11), with the reference state being an ideal solution ($\gamma^{\ddagger} = \gamma_A = \gamma_B = 1$), yields

$$k_{actual} = k_{ref} \frac{\gamma_A \gamma_B}{\gamma^{\ddagger}}$$

(9-12)

This relation, the Brønsted-Bjerrum equation, is the basis for considering salt effects and solvent effects. The key assumption in its derivation is that the reaction rate is proportional to the *concentration* of the activated complex, $[(AB)^{\ddagger}]$, and not to its activity.

9-4   SOLVENT EFFECTS ON POLAR AND IONIC REACTIONS

We can consider three cases for a second-order reaction: two polar molecules, two ions, and one of each.

Kirkwood[5] gives expressions for the free energy of transfer of a polar molecule (of dipole moment μ and radius r) from a medium of unit dielectric constant to one having a value ϵ:

$$\Delta G = \frac{-N\mu^2}{r^3} \frac{\epsilon - 1}{2\epsilon + 1}$$

(9-13)

The value of the rate constant according to Eq. (9-12) becomes

$$k_{actual} = k_{ref} \exp\left(\frac{N}{RT}\right) \frac{\epsilon - 1}{2\epsilon + 1} \left(\frac{\mu_{\ddagger}^2}{r_{\ddagger}^3} - \frac{\mu_A^2}{r_A^3} - \frac{\mu_B^2}{r_B^3}\right) \qquad (9\text{-}14)$$

where k_{ref} refers to the state with $\epsilon = 1$ (as in the gas phase). The data given in Table 9-1 for reaction (9-2) are shown in Fig. 9-1 in a plot of log k versus $(\epsilon - 1)/(2\epsilon + 1)$ in accord with relation (9-14) provided the μ^2/r^3 terms are solvent-independent. Perhaps we should limit our interpretation to one qualitative matter. The positive slope of Fig. 9-1 is consistent with development of ionic charge in the activated complex for which one would predict a large value of $\mu_{\ddagger}$ and a small value of $r_{\ddagger}$. It can also be shown (Prob. 9-1) that a linear variation of ln k with $1/\epsilon$ is expected to a reasonable accuracy.

An electrostatic expression[5] for the free energy of transfer of two ions of charges z_A and z_B from infinite separation to their reaction distance is

$$\Delta G_{el}^{\ddagger} = \frac{N z_A z_B e^2}{\epsilon r_{\ddagger}} \qquad (9\text{-}15)$$

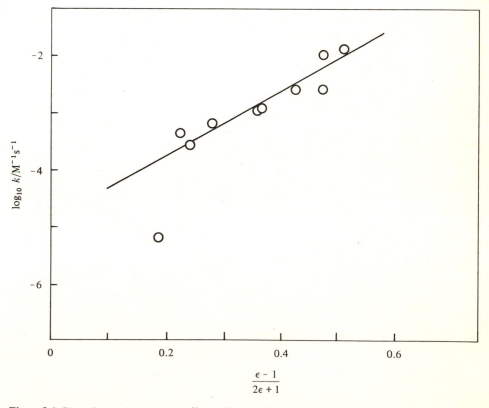

Figure 9-1 Plot of $\log_{10} k$ versus $\epsilon - 1/2\epsilon + 1$ for reaction 9-2. See Eq. (9-14) and Table 9-1.

The same procedure results in the expression

$$k_{\text{actual}} = k_{\text{ref}} \exp\left(\frac{-Nz_A z_B e^2}{\epsilon R T r_{\ddagger}}\right) \tag{9-16}$$

where the reference state has $\epsilon = \infty$. A linear variation of $\ln k$ with $1/\epsilon$ is predicted.

A similar treatment for reaction of an ion and a polar molecule gives

$$k_{\text{actual}} = k_{\text{ref}} \exp\left(\frac{Nz^2 e^2}{2\epsilon R T}\right)\left(\frac{1}{r_{\text{ion}}} - \frac{1}{r_{\ddagger}}\right) \tag{9-17}$$

Thus in all three cases examined, the variation of rate constant with dielectric constant is given by the linear plot of $\ln k$ versus $1/\epsilon$.

9-5 SALT EFFECTS ON SECOND-ORDER IONIC REACTIONS

The equation of Debye and Hückel for the activity coefficient of an ion of charge z_i is[†]

$$\log_{10} \gamma_i = \frac{-A z_i^2 \mu^{1/2}}{1 + \mu^{1/2}} \tag{9-18}$$

where μ is the ionic strength of the solution ($\mu = \frac{1}{2}\Sigma\,[i]\,z_i^2$, the sum being made for all ions present) and A is a collection of physical constants with the value 0.509 for water at 298 K. (Values in water at other temperatures are tabulated.[8])

Substitution of Eq. (9-18) into (9-12), noting that $z^{\ddagger} = z_A + z_B$, yields for a reaction following second-order kinetics

$$\log k = \log k_{\text{ref}} + \frac{A(z_A + z_B)^2 \mu^{1/2}}{1 + \mu^{1/2}} - \frac{A z_A^2 \mu^{1/2}}{1 + \mu^{1/2}} - \frac{A z_B^2 \mu^{1/2}}{1 + \mu^{1/2}} \tag{9-19}$$

$$\log k = \log k_{\text{ref}} + 2 z_A z_B \frac{A \mu^{1/2}}{1 + \mu^{1/2}} \tag{9-20}$$

Taking the reference state as pure water, $\mu = 0$, Eq. (9-20) predicts that a plot of $\log k$ versus $2A\mu^{1/2}/(1 + \mu^{1/2})$ should have a slope given as the product of ionic charges on the reactants. Thus the rate of a second-order reaction between two cations or two anions should increase with increasing ionic strength, whereas the rate of a reaction between oppositely charged ions should decrease.

The validity of Eq. (9-20) has been proved numerous times, but one must not

[†]Other forms of Eq. (9-18) are sometimes given. One is the so-called Debye-Hückel "limiting law,"

$$\log_{10} \gamma_i = -A z_i^2 \mu^{1/2} \tag{9-18a}$$

which is valid only at low ionic strength. Other forms with a nonunity coefficient of $\mu^{1/2}$ in the denominator and/or an added term linear in μ are sometimes used. For our purposes (9-18) or often even (9-18a) will suffice, although the reader should consult detailed references[6,7] as needed. The nonunity coefficient can be written as βa_i, where a_i is the distance of closest approach of the two ions and β a physical constant (in water 0.33×10^{-10} m^{-1}). In view of the limitations of the theory it is doubtful whether great significance can be attributed to a distance a_i so calculated.

expect it to be any more accurate than the Debye-Hückel relation itself. It cannot, therefore, be applied at high ionic strength, especially if the reactants bear high ionic charges.

Some authors object to the use of Eq. (9-18) in the form given. Their reason is that it contains a quantity, the single-ion activity coefficient, which cannot be measured directly. Identical results are obtained if the activity coefficients are expressed in terms of the values for salts. The derivations, however, are not given here.[9]

Little new information is gleaned from a study of salt effects in the dilute solution range in a study involving properly characterized reactants. The magnitude of the effect is based entirely upon the ionic charges, which are known in most instances. One use of salt effect studies has, in fact, been to determine or confirm the ionic charge of one reactant. Both $e^-(aq)$[10] and $Cu^+(aq)$[11] have been examined from that point of view, as have certain metalloproteins.[12] An illustration of the principles can be made by using published data[11] on the reaction of Cu^+ and $Co(NH_3)_5Cl^{2+}$. Table 9-3 summarizes the kinetic data on the reaction. Figure 9-2 depicts the variation of $\log k$ with $\mu^{1/2}$ [Eq. (9-18a)] and with $\mu^{1/2}/(1 + \mu^{1/2})$ [Eq. (9-18)]. The slope shown for the linear portion at low ionic strength is the theoretical value (2.04) for a second-order reaction between ions of charge 1+ and 2+.

One must make allowance for salt effects when studying the kinetics of a reaction in which the reactants are ions. Imagine, for example, that the following reaction was studied with these starting concentrations: 2.00×10^{-3} M $[Co(NH_3)_5OH_2](ClO_4)_3$ and 6.00×10^{-3} M HCl.

$$Co(NH_3)_5OH_2^{3+} + Cl^- \longrightarrow Co(NH_3)_5Cl^{2+} + H_2O$$

The ionic strength of the starting solution would be 0.018 mol dm^{-3}, whereas that at the end, assuming the reaction went to completion, would be 0.010 mol dm^{-3}. The second-order rate constant for the reaction would increase steadily throughout the run owing to the change in μ. According to Eq. (9-20), the increase would amount to 12 percent (i.e., the value of k near completion would be some 1.12 times that mea-

Table 9-3 Ionic strength effects on the reaction[a] of $Cu^+(aq)$ and $Co(NH_3)_5Cl^{2+}$

Ionic strength μ/M	$10^{-4}\ k$/M^{-1} s^{-1}	$\mu^{1/2}$	$\dfrac{\mu^{1/2}}{1 + \mu^{1/2}}$
9.4×10^{-4}	1.50	0.0307	0.0297
4.23×10^{-3}	1.84	0.065	0.061
9.8×10^{-3}	2.21	0.099	0.090
2.25×10^{-2}	2.86	0.150	0.130
3.98×10^{-2}	2.81	0.199	0.166
6.25×10^{-2}	4.08	0.250	0.200
1.00×10^{-1}	3.79	0.316	0.240
2.00×10^{-1}	4.98	0.447	0.309
1.00	7.60	1.00	0.500
3.00	11.3	1.73	0.634

[a]At 25.0°C; data from Ref. 11.

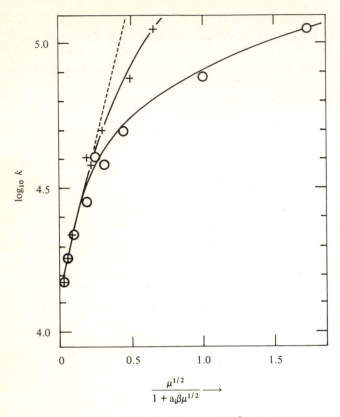

Figure 9-2 Plot of log k for the reaction of $Cu^+(aq)$ with $Co(NH_3)_5Cl^{2+}$ (data from Table 9-3 and Ref. 11). The abscissa represents either $\mu^{1/2}$ [circles, Eq. (9-18a), $a_i\beta = 0$] or $\mu^{1/2}/(1 + \mu^{1/2})$ [crosses, Eq. (9-18a), $a_i\beta = 1.00$]. The slope of the linear portion at low ionic strength is the theorectical value for this reaction, $2Az_A z_B = 2.04$.

sured initially). Worse yet, still greater discrepancies would be found from run to run employing different initial concentrations of the reactants.

The problem is usually circumvented by adding "inert" electrolytes to maintain constant ionic strength between runs. In this instance, for example, one might add perchloric acid, perchlorate ion being largely noncomplexing in water, such that each run had an ionic strength of, say, 0.500 mol dm^{-3}. Thus in the first experiment 0.482 M $HClO_4$ would be added. The final ionic strength would be 0.492 M, which means that k would increase only an insignificant 0.9 percent throughout the run [assuming Eq. (9-20) remains valid at the higher ionic strength]. Moreover, by adjusting the perchloric acid concentration so that each run has $\mu = 0.500$ mol dm^{-3}, values of k from experiments at different reactant concentrations should agree, provided the rate is independent of $[H^+]$. This notion, known as the *constant ionic strength principle*, is not exact in that individual electrolytes do not have identical activity coefficients at the same concentration, particularly at higher ionic strengths where specific interactions may become more important. It is a difficult problem to decide whether a small but systematic variation of the rate constant with reagent concentration is due to the growing importance of a new and genuine reaction pathway or to a medium effect arising from failure of the constant ionic strength principle over a wide range of concentrations.

9-6 SALT EFFECTS ON OTHER REACTION ORDERS

Consideration of the ionic strength dependence of a rate constant of other than second-order form is in order. A first-order reaction will show negligible salt effects because reactant and activated complex bear the same ionic charge; the same thing is true of a second-order reaction in which one component is a neutral molecule, including solvent.[†]

The question of the general expression for salt effects can best be considered by use of examples. Consider the reaction of Fe(III) and Sn(II) in $HClO_4/HCl$ solutions in which the major species are $Fe^{3+}(aq)$ and $Sn^{2+}(aq)$. The kinetic expression contains these terms:

$$\frac{-d[Fe^{3+}]}{dt} = k_1 \frac{[Fe^{3+}][Sn^{2+}]}{[H^+]} + k_2[Fe^{3+}][Sn^{2+}][Cl^-] \qquad (9\text{-}21)$$

Both k_1 and k_2 should show an appreciable salt effect, and we shall endeavor to develop a general method. One defines a *net activation process*[13] as a hypothetical reaction in which the activated complex is formed directly from the predominant forms of the reactants. For the two terms in the rate law given, the net activation reactions are

$$Fe^{3+} + Sn^{2+} + H_2O \overset{K_1^{\ddagger}}{=\!=\!=} [FeSnOH^{4+}]^{\ddagger} + H^+ \qquad (\Delta z_1{}^2)^{\ddagger} = 4 \qquad (9\text{-}22)$$

$$Fe^{3+} + Sn^{2+} + Cl^- \overset{K_2^{\ddagger}}{=\!=\!=} [FeSnCl^{4+}]^{\ddagger} \qquad (\Delta z_2{}^2)^{\ddagger} = 2 \qquad (9\text{-}23)$$

By the Debye-Hückel equation, the concentration of each activated complex (and therefore the rate) will vary with ionic strength in proportion to the values of the hypothetical constants $K_1^{\ddagger}$ and $K_2^{\ddagger}$.

The salt effects on k_1 and k_2 take the forms

$$\log k_1 = \log k_1^{\circ} + \log \frac{\gamma_{Fe^{3+}}\gamma_{Sn^{2+}}}{\gamma_{H^+}\gamma_{\ddagger}}$$

$$\log k_1 = \log k_1^{\circ} + \frac{4A\mu^{1/2}}{1 + \mu^{1/2}} \qquad (9\text{-}24)$$

$$\log k_2 = \log k_2^{\circ} + \log \frac{\gamma_{Fe^{3+}}\gamma_{Sn^{2+}}\gamma_{Cl^-}}{\gamma_{\ddagger}} \qquad (9\text{-}25)$$

[†]Actually, neutral molecules have activity coefficients which are linear functions of ionic strength. The magnitude is much smaller than for ions, however, and it can be neglected in the present context.

$$\log k_2 = \log k_2^\circ + \frac{2A\mu^{1/2}}{1 + \mu^{1/2}} \tag{9-26}$$

The magnitude of the salt effect is given by the value and sign of the coefficient of the quantity $A\mu^{1/2}/(1 + \mu^{1/2})$. The value for k_1 is 4 [Eq. (9-24)], and for k_2 it is 2 [Eq. (9-26)]. Those are just the respective values of $(\Delta z^2)^\ddagger$ shown for the net activation reactions of Eqs. (9-22) and (9-23), and that can be shown to be a general result. For the general situation in which the net activation process is

$$A^{z_A} + B^{z_B} + C^{z_C} + \cdots = (X^{z_X})^\ddagger + Y^{z_Y} + \cdots$$

$$\log k = \log k^\circ + \log \frac{\gamma_A \gamma_B \gamma_C \cdots}{\gamma_\ddagger \gamma_Y \cdots} \tag{9-27}$$

which becomes, by virtue of charge conservation ($z_A + z_B + z_C + \cdots = z_X + z_Y + \cdots$),

$$\log k = \log k^\circ + \frac{(\Delta z^2)^\ddagger A\mu^{1/2}}{1 + \mu^{1/2}} \tag{9-28}$$

where $(\Delta z^2)^\ddagger = z_X^2 + z_Y^2 - z_A^2 - z_B^2 - z_C^2$. A further illustration of these effects is found in data for the reaction[14] of V^{3+} and VO_2^+, which follows the rate law

$$\frac{-d[V^{3+}]}{dt} = k \frac{[V^{3+}][VO_2^+]}{[H^+]} \tag{9-29}$$

The values of k as a function of ionic strength are:

μ/M	0.060	0.20	0.50	1.0	2.0	2.5
k/s^{-1}	16.0	15.3	14.9	15.4	16.9	17.6

We see that k is largely independent of μ. That is consistent with the value of $(\Delta z^2)^\ddagger$ for the net activation process.

$$V^{3+} + VO_2^+ + H_2O \longrightarrow [V_2O_3H_2^{3+}]^\ddagger + H^+ \quad (\Delta z^2)^\ddagger = 0$$

9-7 SALT EFFECTS AND REACTION MECHANISMS

This section is meant to convey just one point: *The magnitude of the salt effect is a property of the rate law, not of the reaction mechanism.* The classic example[15] of this point is the question whether the urea synthesis from ammonium cyanate is a reaction between the ions NH_4^+ and NCO^- or the neutral molecules NH_3 and $HNCO$. The distinction proves quite impossible on such a basis.[†]

   The observed salt effect on a rate constant which is a composite of true rate constants and equilibrium constants is the net effect from both sources, the so-called

---

[†]Although considering the bonds which need to be broken and made and the structure of the requisite transition state, one is quite likely to opt for the latter.

"primary salt effect" on the former and the "secondary salt effect" on the latter. (The two have comparable magnitudes and forms, so the terminology may be misleading.)

We shall illustrate the key thesis with one example. Consider the following two mechanisms, schemes I and II, to account for the k_2 rate term of Eq. (9-21).

Scheme I:
$$Fe^{3+} + Cl^- \underset{}{\overset{K_{Fe}}{\rightleftharpoons}} FeCl^{2+} \qquad \Delta Z^2 = -6 \tag{9-30}$$

$$FeCl^{2+} + Sn^{2+} \xrightarrow{k_3} \text{products} \qquad (\Delta Z^2)^{\ddagger} = +8 \tag{9-31}$$

Scheme II:
$$Sn^{2+} + Cl^- \underset{}{\overset{K_{Sn}}{\rightleftharpoons}} SnCl^+ \qquad \Delta Z^2 = -4 \tag{9-32}$$

$$SnCl^+ + Fe^{3+} \xrightarrow{k_4} \text{products} \qquad (\Delta Z^2)^{\ddagger} = +6 \tag{9-33}$$

The rate constant k_2, either $k_3 K_{Fe}$ or $k_4 K_{Sn}$, has a salt effect which is the same regardless of mechanism. In scheme I, the value of $\log (k_3/k_3^{\circ}) = 8A\mu^{1/2}/(1 + \mu^{1/2})$, and $\log (K_{Fe}/K_{Fe}^{\circ}) = -6A\mu^{1/2}/(1 + \mu^{1/2})$, whereas the corresponding parameters in scheme II are $\log (k_4/k_4^{\circ}) = -4A\mu^{1/2}/(1 + \mu^{1/2})$ and $\log (K_{Sn}/K_{Sn}^{\circ}) = 6A\mu^{1/2}/(1 + \mu^{1/2})$; in either case the net effect is $+2A\mu^{1/2}/(1 + \mu^{1/2})$, exactly as shown by Eq. (9-27). Any other mechanism which would lead to the same rate law would be characterized by the same ionic strength dependence. As a consequence, salt effects cannot be invoked as the basis for distinguishing between alternatives consistent with other kinetic data.

9-8 INFLUENCE OF PRESSURE ON SOLUTION REACTIONS

The thermodynamic relation

$$\left(\frac{\partial \Delta G^{\circ}}{\partial P}\right)_T = \Delta V^{\circ} \tag{9-34}$$

applied to $\Delta G^{\ddagger}$, with $k = (RT/Nh) \exp (-\Delta G^{\ddagger}/RT)$, gives

$$\left(\frac{\partial \ln k}{\partial P}\right)_T = \frac{\partial (-\Delta G^{\ddagger}/RT)}{\partial P} = -\frac{\Delta V^{\ddagger}}{RT} \tag{9-35}$$

and a means of measuring the *volume of activation*. Integration[†] of Eq. (9-35) affords an expression

$$\ln k = \ln k^{\circ} - \frac{\Delta V^{\ddagger}}{RT} P \tag{9-36}$$

This relation indicates that the rate constant increases with increasing pressure if $\Delta V^{\ddagger}$ is negative (i.e., if the activated complex has a smaller molar volume than the reactants

---

[†]The integration is done under the assumption that the volume of activation is independent of pressure. If that is not the case, a further quantity termed the compressibility is introduced to give the expressions:

$$\Delta \beta^{\ddagger} = -d \, \Delta V^{\ddagger}/dP \qquad \ln k = a + bP + cP^2 \qquad \Delta V_0^{\ddagger} = -bRT \qquad \Delta \beta^{\ddagger} = 2cRT$$

together). In practice, pressures of the order of 0.5 to 5 kbar are needed if values are to be meaningful. A plot of $\ln k$ versus p is made; it should, according to Eq. (9-36), be linear and have a slope equal to $-\Delta V^{\ddagger}/RT$.

If we consider, for example, a reaction having $\Delta V^{\ddagger} = +5.0$ cm^3 mol^{-1} at 298 K, a reasonably typical value, the pressure needed to change the rate by a factor of 2 from its value at atmospheric pressure (1 bar $= 10^5$ N m$^{-2} = 10^5$ Pa $= 0.1$ MPa) can be calculated as follows:

$$\ln \frac{k_P}{k^\circ} = \ln \frac{1}{2} = \frac{-5.0 \times 10^{-6} \text{ m}^3 \text{ mol}^{-1}}{8.31 \text{ J mol}^{-1} \text{ K}^{-1} \times 298 \text{ K}} (P - 10^5)$$

$$P = 10^5 + \frac{\ln 2 \times 8.31 \times 298}{5.0 \times 10^{-6}} = 3.4 \times 10^8 \text{ Pa}$$

which amounts to 3.4×10^3 bar. Even when the volume of activation is relatively large, $\Delta V^{\ddagger} = \pm 40$ cm^3 mol^{-1}, a pressure of 430 bar is needed to cause a factor of 2 change in rate. That stands in contrast to the relatively small variation in temperature needed to effect a comparable change. As a consequence, values of $\Delta H^{\ddagger}$ and $\Delta S^{\ddagger}$ are routinely determined, whereas values of $\Delta V^{\ddagger}$ are much less frequently reported. The special equipment needed to obtain pressures of thousands of atmospheres has limited these studies to a relatively few laboratories. That is unfortunate, because the value of $\Delta V^{\ddagger}$, which represents the change in volume of the reacting system in passing to the transition state, is a meaningful parameter. The corresponding value of $\Delta S^{\ddagger}$ is often given a similar interpretation to infer details of structural changes during the activation step. Yet, volume being a more easily appreciated quantity than entropy, values of $\Delta V^{\ddagger}$ can be of considerable value.

One study[16] useful for illustrative purposes is the thermal decomposition of di-t-butyl peroxide. The data given in Table 9-4 are plotted as suggested by Eq. (9-36). The plot of $\ln k$ versus P is shown in Fig. 9-3, and the slope, -1.64×10^{-9} m s^2 kg^{-1}, gives $\Delta V^{\ddagger} = 5.4$ cm^3 mol^{-1}:

$$\Delta V^{\ddagger} = -\text{slope} \times RT = 1.64 \times 10^{-9} \text{ m s}^2 \text{ kg}^{-1} \times \frac{8.314 \text{ kg m}^2 \text{ s}^{-2}}{\text{mol K}} \times 393 \text{ K}$$

$$= 5.4 \times 10^{-6} \text{ m}^3 \text{ mol}^{-1} = 5.4 \text{ cm}^3 \text{ mol}^{-1}$$

The positive value of $\Delta V^{\ddagger}$ is consistent with a mechanism in which a dissociative rate-limiting step is proposed. As in the study of temperature dependences, one must

Table 9-4 Pressure dependence of the rate constant
for the thermal decomposition of Di-t-butyl peroxide[a]

P/MPa	$10^6 \, k/\text{s}^{-1}$	P/MPa	$10^6 \, k/\text{s}^{-1}$
0.1	13.4	448	6.6
204	9.5	527	5.7
290	8.0		

[a]In toluene at 120°C. Data from Ref. 16.

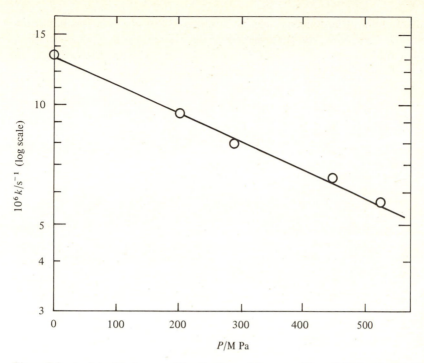

Figure 9-3. A plot of ln k versus P for the thermal decomposition of di-t-butyl peroxide in toluene at 120°C. Data from Table 9-4 and Ref. 16.

be aware that the experimental rate constant will represent a composite of two or more rate constants if the mechanism is at all complex. In such a case the plot of ln k versus P may not be linear, and for the same reason ln k versus $1/T$ will not be. Moreover, considering the values of $\Delta V^{\ddagger}$ in a series of compounds, one recognizes that the relative importance of the rate constants in the composite may differ, so that comparisons must be made with caution.

9-9 CONCENTRATION UNITS IN SOLUTION REACTIONS

The value of $\Delta S^{\ddagger}$ is dependent upon the choice of concentration units for the rate constant; the latter is equivalent to the choice of standard state.

Consider just one case. If a particular second-order rate constant is expressed in the customary units M^{-1} s^{-1} (dm^3 mol^{-1} s^{-1}) and has $\Delta S^{\ddagger} = -21.0$ J mol^{-1} K^{-1}, then factors of $R \ln 10^3$ must be introduced if the concentration scale is changed to a volume different from 1 dm^3 by a factor 10^3. If the units dm^3 $mmol^{-1}$ s^{-1} are used, $\Delta S^{\ddagger} = -21.0 - R \ln 10^3 = -78.4$ J mol^{-1} K^{-1}. Units cm^3 mol^{-1} s^{-1} for k would give $\Delta S^{\ddagger} = -21.0 + R \ln 10^3 = 36.4$ J mol^{-1} K^{-1}. Clearly the inherent value or sign of $\Delta S^{\ddagger}$ is of no intrinsic importance unless the reference state is kept in mind and comparison reactions are expressed on the same basis.

A similar statement has already been made about a second-order gas-phase reaction in Sec. 6-2. First-order reactions are not influenced; for other cases the effect depends upon the order of the reaction. Comparisons of the magnitude or sign $\Delta S^{\ddagger}$ values for reactions of different kinetic orders is clearly unjustified.

REFERENCES

1. Clark, I. D., and R. P. Wayne: in Bamford and Tipper, "Comprehensive Chemical Kinetics," Elsevier Scientific Publishing Co., Inc., New York, vol. 2, p. 302.
2. Noyes, R. M.: *Progr. in Reaction Kinetics*, 1:135, 136 (1961).
3. Ref. 1, pp. 306–308.
4. Lin, S. H., K. P. Li, and H. Eyring: in Eyring, Henderson, and Jost (eds.), "Physical Chemistry, an Advanced Treatise," Academic Press, Inc., New York, 1975, vol. VII, pp. 10–18.
5. Kirkwood, J. G.: *J. Chem. Phys.*, 2:351 (1954); see Frost and Pearson, pp. 135–145.
6. Davies, C. W.: *Progr. in Reaction Kinetics*, 1:161 (1961).
7. Harned, H. S., and B. B. Owen: "The Physical Chemistry of Electrolyte Solutions," 2d ed., Reinhold Book Corporation, New York, 1950.
8. Manov, G. G., R. G. Bates, W. J. Hamer, and S. F. Acree: *J. Am. Chem. Soc.*, 65:1765 (1943).
9. King, E. L.: in P. H. Emmett (ed.), "Catalysis," Reinhold Publishing Corporation, New York, 1955, vol. II, pp. 361–369.
10. Hart, E. J.: *Science*, 146:19 (1964).
11. Parker, O. J., and J. H. Espenson: *J. Am. Chem. Soc.*, 91:1968 (1969).
12. Rosenberg, R. C., S. Wherland, R. A. Holwerda, and H. B. Gray: *J. Am. Chem. Soc.*, 98:6364 (1976).
13. Daugherty, N. A., and T. W. Newton: *J. Phys. Chem.*, 68:612 (1964).
14. Newton, T. W., and S. W. Rabideau: *J. Phys. Chem.*, 63:365 (1959).
15. Frost, A. A., and R. G. Pearson: "Chemical Kinetics," 2d ed., John Wiley & Sons, Inc., New York, 1965, pp. 307–316.
16. Walling, C., and G. Metzger: *J. Am. Chem. Soc.*, 81:5365 (1959).

PROBLEMS

9-1 *Solvent effects.* Equation (9-13) contains the factor $(\epsilon - 1)/(2\epsilon + 1)$.

(*a*) Show that the factor may be expanded as

$$\frac{\epsilon - 1}{2\epsilon + 1} = \frac{1}{2} + \frac{3}{4\epsilon} + \frac{3}{8\epsilon^2} - \frac{3}{16\epsilon^3} + \cdots$$

(*b*) Find the range of dielectric constants over which the first two terms of the expansion hold to within 1 percent; to within 10%.

(*c*) Show that in the range of (*b*) the rate constant depends upon dielectric constant according to

$$\ln \frac{k_{\text{actual}}}{k_\infty} = \frac{3N}{4\epsilon RT} \left[\frac{A^2}{r_A^3} + \frac{B^2}{r_B^3} - \frac{\ddagger^2}{r_\ddagger^3} \right]$$

where k_∞ represents the rate constant in a medium of infinite dielectric constant. Test this expression by a plot of $\ln k$ versus $1/\epsilon$ using data in Table 9-1 for the reaction of Eq. (9-2).

9-2 *Salt effects.* Rosseinsky [*J. Inorg. Nucl. Chem.*, 33:3976 (1971)] pointed out that the rate of reaction of vanadium(V) with iodide ions is independent of ionic strength. The rate law is

$-d[V(V)]/dt = k[V(V)][I^-][H^+]^2$. What is the charge on the predominant V(V) species in these solutions?

9-3 *Salt effects.* Postmus and King [*J. Phys. Chem.*, **59**:1216 (1955)] found that the reaction $Cr^{3+} + NCS^- = CrNCS^{2+}$ follows this rate expression in the forward direction:

$$\frac{d[CrNCS^{2+}]}{dt} = (k_1 + k_2[H^+]^{-1})[Cr^{3+}][NCS^-]$$

At 25°C and $\mu = 0.070$ M, $k_1 = 4.28 \times 10^{-6}$ M^{-1} s^{-1} and $k_2 = 1.96 \times 10^{-8}$ s^{-1}. Estimate k_1 and k_2 at $\mu = 0.010$ M.

9-4 *Salt effects.* Seel and Winkler found that the decomposition of *N*-hydroxylamine-*N*-sulfonate $(=X^{2-})$ follows a rate law $-d[X^{2-}]/dt = k[X^{2-}][H^+]$. The slope of log k versus $0.509 \mu^{1/2}$, in a series of experiments at pH 7.00, has a slope -3.0. Reconcile that result with Eq. (9-20). (Remember that pH $= -\log a_{H^+}$.)

9-5 *Volume of activation.* The rate constant for the base hydrolysis of $ClC(CH_3)_2C\equiv CH$ [LeNoble, Tatsukami, and Morris, *J. Am. Chem. Soc.*, **92**:5681 (1970)] varies with pH as shown below. Compute $\Delta V^{\ddagger}$.

p/kbar	0.001	1.055	3.023	5.062
$10^6 k_{298}$/M^{-1} s^{-1}	2.68	2.38	1.55	1.20

9-6 *Volume of activation.* Swaddle [*Inorg. Chem.*, **15**:2644 (1976)] gives the following values for the rate of exchange of *trans*-Co(en)$_2$(H$_2^{18}$O)$_2^{3+}$ and solvent at 308 K. Calculate $\Delta V^{\ddagger}$. What is the implication of that value for the mechanism?

p/MPa	0.1	6.4	102.5	152.4	201.3	250.0	301.9
$10^5 k_{ex}$/s^{-1}	6.29	6.27	5.05	4.23	4.17	3.52	3.16

Note: 0.1 MPa (megapascal) = 1 bar.

TEN

REACTIONS AT EXTREME RATES

Specialized techniques are required for the determination of reaction rates outside the time frame in which one usually envisages conducting chemical operations. This chapter considers the wide range of methods available. Most interest focuses on fast reactions, rather than extremely slow ones, say those with mean reaction times ranging from roughly 1 min to 10^{-13} s. The latter is the time scale of molecular vibrations. The advances in electronic instrumentation and the invention of some truly novel approaches for rapid phenomena have made it a truism that no reaction is properly labeled "immeasurably rapid."

10-1 SURVEY OF METHODS FOR VERY FAST REACTIONS

The customary methods for "conventional" reaction rates involve first mixing the reactants and then determining the subsequent concentration changes. The latter may be done by a continuous monitoring or by sampling. In the latter method the reaction must be quenched, unless it is very slow, by a chemical reagent or a sudden cooling.[1,2]

Flow methods entail the mixing of separate solutions of two reactants, and they permit the observation of reactions occurring as rapidly as 0.1 ms. (The usual range is 1 ms to 10 s.) The instrumentation (Sec. 10-2) is now commonplace, and it permits routine work in the time frame indicated. It is by far the most widely applicable and general of the methods described.

Grouped under the heading "perturbation methods" or "relaxation methods" are techniques by which an equilibrium is upset by a sudden change in a physical property such as temperature or pressure. The readjustment of the equilibrium concentrations is then measured. The time resolution may be as short as 10^{-10} s, although 10^{-6} s can be considered the more practical limit generally available. The method requires no mixing

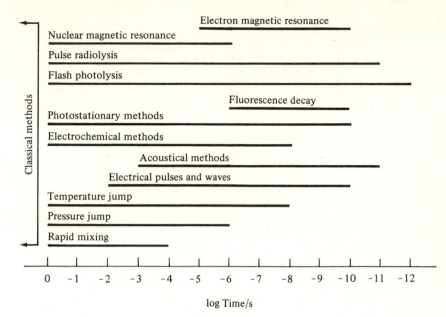

Figure 10-1 Summary of fast reaction techniques and their time ranges of applicability (from Hammes, Ref. 3).

(and that is what permits its time resolution), but it is, of course, applicable only to equilibria which are properly poised under the conditions of the measurements.

Magnetic resonance methods include applications of nmr and epr, and they are especially valuable for rapid exchange reactions. The occurrence of exchange leads to line broadening, and the degree of such broadening depends, among other things, on the reaction rate.

Unstable species can be produced by the absorption of radiation. Flash photolysis and pulse radiolysis techniques are of considerable value in providing a means of directly producing and studying reactive materials with short lifetimes; the techniques permit their production in a very short period of time. That permits direct observation of the reactions of transient intermediates, which is not possible by using conventional techniques.

A summary[3] of the time ranges available by these and similar methods is given in Fig. 10-1. Details of selected methods are given in subsequent sections of this chapter.

10-2 FLOW METHODS FOR RAPID REACTIONS

The most popular method is the *stopped-flow* technique. A schematic diagram of the apparatus is shown in Fig. 10-2. Syringes **A** and **B**, filled through suitable valves, contain the separate reactants. The common drive mechanism **D**, usually an air-driven piston, forces the solution into mixing chamber **M** and then into the stopping syringe **C**, whose plunger comes to rest against fixed stop **S**. That action triggers the recording

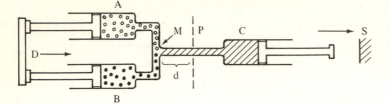

Figure 10-2 Schematic diagram of a stopped-flow instrument whose operation is described in the text.

device, usually an oscilloscope. The reaction is observed in the now-stationary solution at point **p** a distance **d** from the mixing chamber. The method of detection is most often based on UV-visible spectrophotometry.

The stopped-flow method[4] yields perfectly ordinary kinetic data, values of the property P_t as determined at various times after mixing. The values may be read from a Polaroid photograph or may be obtained in digitized form by use of an electronic transient recorder, which permits easy processing by computer techniques.

Two less popular flow methods are the continuous-flow and quenched-flow procedures. In the former method measurements are made at different distances down the flow tube (i.e., at varying distances **d** in Fig. 10-2), the stopping syringe **C** having been eliminated. The principle is that measurements along the flow stream amount in effect to determinations at different times; with a known and constant flow rate, the time to which each reading applies is known. The measurement might be made by positioning several detectors along the tube or moving a single detector to different positions. There is a slight improvement in time resolution in the continuous-flow method as compared with the stopped-flow method, 10^{-4} versus 10^{-3} s.

On the other hand, the continuous-flow method uses large quantities of solutions and for that reason is not widely encountered in practice. The quenched-flow procedure is also a specialized method finding only limited applicability. In this procedure samples of the reaction mixture are rapidly mixed with a quenching mixture, following which the analytical determination can be made at leisure. The timing of different samples can be varied by using different flow rates or different distances along the flow stream. The best application of the method concerns the study of exchange reactions which occur too rapidly for the usual sampling techniques. Since exchange reactions occur without net change, the use of this rapid quenching method permits conventional separation and isotopic assay techniques to be applied. The quenching reagent must be specific for the particular reaction; a sudden pH change or electron transfer reagent might be useful in specific instances. By using the method, the rate of the rapid electron exchange between $Fe(\eta\text{-}C_5H_5)_2$ and $Fe(\eta\text{-}C_5H_5)_2{}^+$ at $-70°C$ in methanol has been evaluated;[5] in these runs, the exchange half-times were 1 to 4 ms.

10-3 RELAXATION METHODS

The basis of the relaxation methods, originally developed by Eigen, is the extreme suddenness, relative to the chemical reaction time, with which the physical perturba-

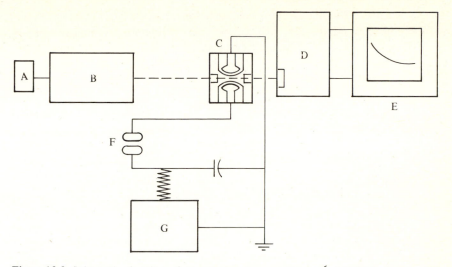

Figure 10-3 Schematic drawing of temperature-jump apparatus.[6] (*A*) Light source; (*B*) mono-chromator; (*C*) observation cell; (*D*) photomultiplier, cathode follower; (*E*) oscilloscope; (*F*) spark gap; (*G*) high voltage.

tion is applied. Most often, a single step-function perturbation is applied; it might be a sudden temperature rise from the discharge of a high-voltage capacitor across an electrolyte solution or the bursting of a diaphragm containing a pressurized solution. The mean reaction time must be longer than the time for the perturbation, and a means of monitoring the ensuing concentration change must be provided. Changes in light absorption or in conductance are often used. Clearly, the method must be a rapidly responding one.

Figure 10-3 shows a schematic diagram[6] of a temperature-jump apparatus based on spectrophotometric detection. The magnitude of the perturbing impulse must be great enough to give a perceptible change in the equilibrium concentration, yet it is highly desirable, as shown below, that the shift be small enough that the rate of approach to the new equilibrium will conform to a first-order process (regardless of the authentic kinetic order).

The necessary temperature or pressure change depends on the respective values of ΔH° and ΔV° for the equilibrium. A typical apparatus might discharge some 10 to 25 kV in 1 cm^3 of solution producing about 45 J in 10^{-6} s, which corresponds to 4.5×10^7 W.[7,8] A temperature change of 5 to 10° might ensue. Owing to the high rate of the reactions to which the method is usually applied, reequilibration of temperature with the surroundings during the run is not significant.

10-4 KINETIC EQUATIONS FOR RELAXATION KINETICS

Suppose that a simple equilibrium such as

$$A \underset{-1}{\overset{1}{\rightleftarrows}} B \tag{10-1}$$

is suddenly displaced. The considerations advanced in Chap. 3 suggest that the rate of approach to the new equilibrium will follow first-order kinetics under all circumstances with $k_{obs} = k_1 + k_{-1}$. It has been customary to refer to the relaxation time τ, the reciprocal of k_{obs}:

$$\tau^{-1} = k_{obs} = k_1 + k_{-1} \tag{10-2}$$

We have already considered (Sec. 3-3) the specific situation applicable to the more complicated reaction

$$A + B \underset{-1}{\overset{1}{\rightleftharpoons}} C \tag{10-3}$$

and have found that the relaxation time is given by

$$\tau^{-1} = k_{obs} = k_{-1} + k_1([A]_{eq} + [B]_{eq}) \tag{10-4}$$

where the concentrations are those prevailing at the new temperature. The validity of this relation depends on there being a perturbation sufficiently small to permit neglect of quadratic and higher-order terms. We shall now examine the general case[9] applicable to any single-step relaxation process, which can be represented as

$$aA + bB \rightleftharpoons dD + eE \tag{10-5}$$

with a rate given by

$$-\frac{1}{a}\frac{d[A]}{dt} = k_f[A]^a[B]^b - k_r[D]^d[E]^e \tag{10-6}$$

If the perturbation of the concentration of A is represented as $a\delta$, then the values of each concentration relative to the equilibrium values are:

$$[A] = [A]_{eq} + a\delta = [A]_{eq}(1 + a\delta[A]_{eq}^{-1})$$
$$[B] = [B]_{eq} + b\delta = [B]_{eq}(1 + b\delta[B]_{eq}^{-1})$$
$$[D] = [D]_{eq} - d\delta = [D]_{eq}(1 - d\delta[D]_{eq}^{-1})$$
$$[E] = [E]_{eq} - e\delta = [E]_{eq}(1 - e\delta[E]_{eq}^{-1})$$

Substitution of these relations into the rate law and imposition of the equilibrium condition whereby $k_f[A]^a[B]^b = k_r[D]^d[E]^e$ and the experimental condition of a small perturbation permitting the neglect of all terms containing powers of δ beyond unity yield the result

$$\tau^{-1} = k_{obs} = \frac{d\ln\delta}{dt} = k_f[A]_{eq}^a[B]_{eq}^b\left(\frac{a^2}{[A]_{eq}} + \frac{b^2}{[B]_{eq}}\right)$$
$$+ k_r[D]_{eq}^d[E]_{eq}^e\left(\frac{d^2}{[D]_{eq}} + \frac{e^2}{[E]_{eq}}\right) \tag{10-7}$$

This expression applied to the example of Eq. (10-4), with $a = b = d = 1$ and $e = 0$, yields the correct value of the relaxation time.

Many chemical systems are not controlled by a single relaxation process. The algebraic treatment of such systems[6,7,10,11] can become laborious. Without considering such systems in detail, we observe that one relaxation time will be observed for each member in a set of coupled reactions. With sufficient variation of concentration even quite complex systems can be characterized. If some of the steps in a coupled scheme are rapid compared with others, considerable simplification results even though the rapid component must be allowed for.

10-5 MAGNETIC RESONANCE METHODS

Consider a solute containing an nmr-active nucleus such as ^1H which may (or may not) undergo chemical exchange with the same nucleus present in a chemically inequivalent environment. The second nucleus may be in the same solute molecule, a different solute, or the solvent. We shall consider the situation in which the rate of exchange increases from a very low to a very high value. Such experiments are often performed by recording the nmr spectrum over a wide range of temperatures, preferably values at which both the high and low exchange rate limits, as well as intermediate values, can be observed.

The simplest case to consider is that in which there are two different solute molecules each with a single proton. If no exchange occurs at some temperature, each proton resonance occurs at the same chemical shift as if the other solute were absent. This case is shown in the lowest panel of Fig. 10-4, and it is referred to as the no- (or very slow) exchange region. As the rate begins to increase, at higher temperature or a low concentration of catalyst, both resonances begin to broaden and the chemical shift separation decreases. That is known as the intermediate exchange region, and it is shown in panel 2 of Fig. 10-4.

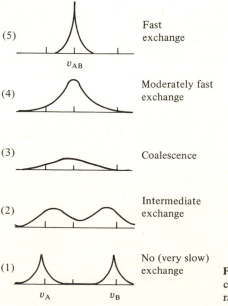

(5)   Fast exchange v_{AB}

(4)   Moderately fast exchange

(3)   Coalescence

(2)   Intermediate exchange

(1)   No (very slow) exchange v_A v_B

Figure 10-4 Proton nmr spectra showing the changes seen as a function of increasing exchange rate (increasing temperature) from (1) to (5).

With a further increase in temperature, the two resonances coalesce to a single broad resonance with a chemical shift centered between the two original sharp peaks. The temperature at which that occurs is known as the coalescence temperature T_c. A precise value is often difficult to determine owing to the gradual change in the spectrum and to its broadness at this point. The rate constant can be estimated from that value alone, although with considerable error. As the temperature is further increased, the spectrum takes on the appearance of a single resonance peak, first yielding a peak broadened by exchange as in panel 4 of Fig. 10-4, and finally a sharp singlet, as in panel 5, of twice the area of the original, separate peaks. The latter corresponds to the region where exchange is now occurring so rapidly that the two sites are, in effect, chemically equivalent on the nmr time scale.

The latter phrase provides the key to the analysis. Since the line broadening due to chemical exchange is usually of the order of 10 to 10^3 radians/s, the nmr method is usually applicable to reactions with lifetimes of the order of 1 to 100 ms. The rate constant may be calculated at T_c, although with considerable uncertainty, from the lifetime. Line shape analysis, usually involving computer simulation, is necessary for accurate determinations, and the most complete description will rely on values from spectra on both sides of T_c.[12,13] The nmr broadening is always described in terms of a first-order rate constant, but, as in any other kinetic determination, one needs to vary the reactant concentrations to determine the dependence of the lifetime (rate constant) on the concentrations of the solutes. The range of rates which can be determined by nmr methods can be extended to much shorter lifetimes by use of special techniques incorporating paramagnetic solutes and the spin-echo experiment.

10-6 FLASH PHOTOLYSIS

Flash photolysis is a most valuable method for the generation of chemical species by a sudden photochemical flash. It is especially useful for the production of energy-rich and reactive species which would not attain detectable levels in thermal reactions. Because of their reactivity, they usually have quite short lifetimes. The problems are those of generating the photochemical transient, detecting and identifying it, and determining its reaction rate with chosen substrates.

The basic apparatus consists of the photochemical initiation unit and the detection system for the transients. The photolyzing light may be provided by high-speed flash lamp or a laser. With the former, the flash duration is usually of the order of microseconds and consists of broad-band radiation which, if desired, may be filtered by special glasses or solutions before it is impinged on the sample. The use of laser sources permits photolysis in 10^{-9} s or less and affords the possibility of still shorter time resolution for reactive transients.

The basic operation consists of an electronic control which triggers the flash. If a flash lamp is used, it is powered by a bank of capacitors. The detecting and analyzing system may be a spectroscopic one, which permits determination of the absorption spectrum of the transient during its decay. Particularly for solution reactions in which most species exhibit wide absorption bands, a much simpler arrangement known as kinetic spectrophotometry is used. In the method the analyzing light is a continuous source with a monochromator set at a wavelength (chosen by experience or trial and error for a given reaction) at which the transient absorbs strongly. After the photolysis lamps are triggered, the absorption of the transient species is recorded as the analyzing light passes through the samples and onto the photomultiplier tube. A monochromator is positioned after the sample. The signal from the photomultiplier tube is displayed on an oscilloscope screen. As with stopped-flow kinetics, the data may be processed from a photograph of the screen or may be digitized for computer treatment.

10-7 PULSE RADIOLYSIS

Pulse radiolysis is a highly specialized technique requiring access to unusual equipment, in particular to an electron accelerator. Although that restricts work to specialized laboratories, the method is nonetheless a highly important one. Pulse radiolysis can be used to generate transient species by direct ionizing radiation, and it has many analogies to the photochemical basis of flash photolysis considered in the preceding section.

If a beam of high-energy electrons impinges on water, a number of stable and unstable species are produced. They arise from the energy transferred to water molecules by the electron beam. It is customary to express the yield of each by a G value, which represents the number of each species formed per 100 eV of energy absorbed by the water. The species formed in the pulse radiolysis of water, with the G value shown as the coefficient of each, are given by the equation

$$4.0\ H_2O \longrightarrow 2.6\ e^-(aq) + 2.6\ HO\cdot + 0.6\ H\cdot + 2.6\ H^+ + 0.4\ H_2 + 0.7\ H_2O_2 \quad (10\text{-}8)$$

The first three are highly reactive entities; they will react with a great many substrates. In some cases it is desired to study the reactions of these species; in others one wishes to use one of the species as a synthetic reagent for the generation of other chemical transients by the use of rapid scavenging reactions. The hydrated electron and the hydrogen atom are strong reducing agents, and the hydroxyl radical $HO\cdot$ is a very powerful oxidant. The respective standard electrode potentials, E°, are -2.7, -2.28 and +1.9 V. In the absence of suitable scavengers the entities will rapidly decay by several reactions including the following:

$$e^-(aq) + H_3O^+ \longrightarrow H\cdot + H_2O \qquad k = 2.2 \times 10^{10}\ M^{-1}\ s^{-1} \qquad (10\text{-}9)$$

$$e^-(aq) + e^-(aq) \longrightarrow H_2 + 2OH^- \qquad k = 5.5 \times 10^9\ M^{-1}\ s^{-1} \qquad (10\text{-}10)$$

$$e^-(aq) + HO\cdot \longrightarrow OH^- \qquad k = 3.0 \times 10^{10}\ M^{-1}\ s^{-1} \qquad (10\text{-}11)$$

$$e^-(aq) + H\cdot \longrightarrow H_2 + OH^- \qquad k = 2.3 \times 10^{10}\ M^{-1}\ s^{-1} \qquad (10\text{-}12)$$

$$HO\cdot + HO\cdot \longrightarrow H_2O_2 \qquad k = 1.0 \times 10^{10}\ M^{-1}\ s^{-1} \qquad (10\text{-}13)$$

Owing to those and several similar reactions, selective methodologies have been developed to permit study of each of $e^-(aq)$, $HO\cdot$, and $H\cdot$ separately. For example, if one wishes to study only $e^-(aq)$ or use exclusively hydrated electrons to generate other species, it is necessary to work in neutral or basic solution to avoid destruction of $e^-(aq)$ by H_3O^+ [Eq. (10-9)]. Hydroxyl radicals and hydrogen atoms are often removed by rapid scavenging reactions such as those with formate ions or *tert*-butanol:

$$HO\cdot \text{ (or } H\cdot) + HCOO^- \longrightarrow H_2O \text{ (or } H_2) + CO_2^- \qquad (10\text{-}14)$$

$$HO\cdot \text{ (or } H\cdot) + (CH_3)_3COH \longrightarrow H_2O \text{ (or } H_2) + \cdot CH_2C(CH_3)_2OH \qquad (10\text{-}15)$$

and hydrogen atoms in alkaline solution are converted into hydrated electrons:

$$H\cdot + OH^- \longrightarrow e^-(aq) + H_2O \qquad k = 2.0 \times 10^7 \text{ M}^{-1} \text{ s}^{-1} \qquad (10\text{-}16)$$

The product radicals may themselves enter into further reaction, but they usually do not interfere because they are generally much less reactive than $e^-(aq)$ and also because they are reducing agents and not oxidizing agents like $HO\cdot$.

Solutions containing hydroxyl radical as the only important energetic species can be prepared by scavenging $e^-(aq)$ with H_3O^+ [Eq. (10-9)] or by using N_2O-saturated solutions. The latter reagent is an effective scavenger of $e^-(aq)$, and in addition it doubles the yield of the desired hydroxyl radical:

$$e^-(aq) + N_2O \xrightarrow{H_2O} N_2 + OH^- + HO\cdot \qquad k = 5.6 \times 10^9 \text{ M}^{-1} \text{ s}^{-1} \qquad (10\text{-}17)$$

The reactivity of the transient of interest may be detected (usually on a microsecond time scale and more rapidly in special circumstances), or one may follow the product of its reaction with time. Quite often relative rates are determined from the yields of different products in competition experiments, and absolute values are available once any single reaction rate is determined. The methods hold considerable potential for the study of reactive organic and inorganic substances.[14,15]

REFERENCES

1. Campion, R. J., C. F. Deck, P. King, Jr., and A. C. Wahl: *Inorg. Chem.*, **6**:672 (1967).
2. Stranks, D. R.: *Discussions Faraday Soc.*, **29**:73 (1960).
3. Hammes, G.: "Investigations of Rates and Mechanisms of Reactions," vol. VI, part II of the series "Techniques of Chemistry," Wiley-Interscience, New York, 1974, p. 4.
4. Chance, B.: in Ref. 3, pp. 6-62.
5. Stranks, D. R.: *Discussions Faraday Soc.*, **29**:73 (1960).
6. Hammes, G. C.: "Principles of Chemical Kinetics," Academic Press, Inc., New York, 1978, p. 189.
7. Hague, D. N.: in C. N. Bamford and C. F. H. Tipper: "Comprehensive Chemical Kinetics," Elsevier Scientific Publishing Co., Inc., New York, vol. I, pp. 112–179.
8. Hammes, G. G., W. Knoche, L. DeMeyer, and A. Persoons: in Ref. 3, pp. 147–236.
9. King, E. L.: *J. Chem. Educ.*, **56**:580 (1979).
10. Eigen, M., and L. DeMaeyer: in Ref. 3, pp. 63–146.
11. Strehlow, H., and W. Knoche: "Fundamentals of Chemical Relaxation," Verlag Chemie, Weinheim, 1977.
12. Swift, T. J.: in Ref. 3, pp. 521–564.
13. Drago, R. S.: "Physical Methods in Chemistry," W. B. Saunders Company, Philadelphia, 1977, pp. 255–262.

14. Buxton, G. V., and R. M. Sellers: *Coord. Chem. Reviews*, **22**:195 (1977).
15. Meyerstein, D.: *Accounts Chem. Research*, **11**:43 (1978).

PROBLEMS

10-1 *Relaxation kinetics in single-stage reactions.* Derive expressions for the reciprocal relaxation times of the following single-stage reactions

(1)

$$A + B \underset{-1}{\overset{1}{\rightleftharpoons}} C + D$$

(2)

$$2A \underset{-2}{\overset{2}{\rightleftharpoons}} A_2$$

10-2 *Relaxation kinetics with a reaction intermediate.* Show that the kinetic scheme with a stationary-state intermediate AB

$$A + B \underset{-1}{\overset{1}{\rightleftharpoons}} AB$$

$$AB \underset{-2}{\overset{2}{\rightleftharpoons}} C$$

corresponds to a single relaxation time

$$\frac{1}{\tau} = k_1 p([A_\infty] + [B]_\infty) + k_{-2}(1 - p)$$

where $p = k_2/(k_{-1} + k_2)$.

10-3 *Relaxation kinetics.* Show that Eq. (10-4) can be transformed to

$$\left(\frac{1}{\tau}\right)^2 = k_1 \Delta^2 + 2k_1 k_{-1} S + k_{-1}^2$$

where $\Delta = [A]_0 - [B]_0$ and $S = [A]_0 + [B]_0$. Consider two series of runs (1) with varying $[A]_0$ and $[B]_0$ such that $[A]_0 = [B]_0$ and (2) keeping S constant and varying Δ. Show the significance in the first case of a plot of $1/\tau^2$ versus $[A]_0$ and in the second of $1/\tau^2$ versus Δ^2.

10-4 *Temperature-jump kinetics.* Silber, Farina, and Swinehart, *Inorg. Chem.*, **8**:819 (1969), have studied the kinetics of complexation of lutetium(III) with anthralinate ion using a temperature-jump method.

If the mechanism is

$$Lu^{3+} + An^- \underset{k_r}{\overset{k_f}{\rightleftharpoons}} LuAn^{2+}$$

show that the relaxation time should be given by

$$\frac{1}{\tau} = k_r + k_f([Lu^{3+}] + [An^-])$$

Given the following data at 12.5°C, carry out an analysis to obtain the individual rate constants.

$10^4 [Lu]_T$	0.83	1.65	3.30	4.94	6.50	8.25	12.3	12.3
$10^4 [An]_T$	0.22	1.22	0.22	0.22	0.22	0.22	0.44	0.22
$10^{-3} \tau^{-1}/s^{-1}$	51	65	84	90	107	123	160	158

10-5 *Hydroxyl radicals.* The acid ionization constant of the short-lived HO· transient is difficult to determine by conventional methods, but an estimate can be made because HO·, but not its conjugate base, O⁻, oxidizes ferrocyanide ions: $HO\cdot + Fe(CN)_6^{4-} \longrightarrow OH^- + Fe(CN)_6^{3-}$. Use the following kinetic data giving the apparent second-order rate constant as a function of pH, from J. Rabani and M. S. Matheson, *J. Am. Chem. Soc.*, 86:3175 (1964), to estimate K_a for the acid dissociation equilibrium $HO\cdot + H_2O \rightleftharpoons H_3O^+ + O^-$.

pH	Neutral	11.94	12.10	12.57	13.07
$10^{-10}\, k_{app}/M^{-1}\, s^{-1}$	1.2	0.49	0.36	0.19	0.06

ELEVEN

EXTRAKINETIC PROBES OF MECHANISM

The primary emphasis prior to this has been placed on the mechanistic information which can be derived or inferred from the kinetics of a single chemical reaction system. Although reaction kinetics remains the single most powerful method for exploring mechanism, no account is complete without reference to complementary techniques. This concluding chapter consists of a potpourri of such topics.

One broad subdiscipline is grouped under the heading linear free-energy relations (LFER). The methods consist of correlations of the experimentally observed changes in reaction rate (or free energy of activation $\Delta G^{\ddagger}$) with changes in molecular structure. Included among the methods are the Hammett and Taft relations of organic chemistry, Marcus's equation for electron transfer, and the Brønsted-Pedersen equation for acid-base catalysis.

Another consideration, and one which underlies the earlier considerations as well, is the structure of the activated complex, not just its elemental composition but also the spatial arrangements of atoms and groups. A study of the stereochemical changes accompanying a particular reaction can provide the key to answering such questions.

The structures of products, the position of a chemically or isotopically tagged grouping, and the detection and/or trapping of a transient intermediate add further insight. (The methods are implicit in earlier material, and they will not be reexamined here.) Finally, we mention the use of isotope rate effects to probe which chemical bonds play a major role in the activation process.

11-1 LINEAR FREE-ENERGY CORRELATIONS

Consider the familiar equations for a rate constant and an equilibrium constant in terms of their respective values of the free energy of activation and the standard free energy of reaction:

$$k = \frac{RT}{Nh} \exp\left(-\frac{\Delta G^{\ddagger}}{RT}\right) \tag{11-1}$$

$$K = \exp\left(-\frac{\Delta G^{\circ}}{RT}\right) \tag{11-2}$$

It is sometimes observed empirically that a closely related group of reactions have rate constants which correlate with other parameters. The correlation is usually a logarithmic one in which $\log k_i$ varies linearly with either (1) $\log K_i$, the equilibrium constants for the series of reactions, or (2) $\log k_i'$, the rate constant for the same set of compounds with a different reagent. Expressed algebraically, such relations can be written as shown in Eqs. (11-3) and (11-4):

$$\log k_i = m \log K_i + b \tag{11-3}$$

$$\log k_i = m' \log k_i' + b' \tag{11-4}$$

Substitution in terms of the values of $\Delta G^{\ddagger}$ and ΔG° transforms Eqs. (11-3) and (11-4) to the following:

$$\Delta G^{\ddagger} = m(\Delta G^{\circ}) + C \tag{11-5}$$

$$\Delta G^{\ddagger} = m'(\Delta G^{\ddagger\prime}) + C \tag{11-6}$$

The form of these equations accounts for the title given to this treatment.

Table 11-1 Tabulation of some LFER correlations

Originator	Type of reaction	Parameter used to correlate variation in rate constant
1. Brønsted-Pedersen	Acid- and base-catalyzed reactions	Ionization constant of the catalyst
2. Grunwald-Weinstein	Solvolysis reactions in mixed solvents	Parameter Y derived from a reference reaction, solvolysis of t-butyl chloride
3. Hammett	Various reactions to probe electronic effects of substituents on aromatic rings	Parameter σ derived from values of K_a for substituted benzoic acids
4. Langford	Metal-ligand complex formation and dissociation	Equilibrium constant for complex formation
5. Marcus	Bimolecular electron transfer, especially outer-sphere reactions	The rates of self-exchange of each partner and the overall equilibrium constants
6. Swain-Scott	Nucleophilic substitution	The rate of a reference reaction, $CH_3Br + OH^-$
7. Taft	Substituent effects in aliphatic systems	Parameter σ^* based on the rates of hydrolysis of alkyl esters

The uses to which such relations can be put are many, and the following are some examples: (1) simply to demonstrate that reactivity correlates with an independent but related parameter, (2) to establish that related reactions have common mechanisms in that a correlation, according to Eq. (11-3) or (11-5), is successful (or to show that they do not have common mechanisms, by the failure of such correlation), (3) to separate and quantify the effects of different variables affecting the same reaction, such as steric and electronic effects, by selecting series in which only one parameter varies, (4) to establish a scale upon which to base the assessment of the importance of similar effects in other series of reactions, including reactions unstudied or unknown at that time. Some of the LFER are listed in Table 11-1, and in subsequent sections certain of them will be examined in greater detail.

One further general comment is pertinent, and it relates to the choice of the kinetic parameter useful for LFER. Given a choice, in the model being considered, between a correlation of $\Delta G^{\ddagger}$ or $\Delta H^{\ddagger}$ with some appropriately chosen parameter, it should be recognized that $\Delta G^{\ddagger}$ is known to about ± 0.13 kJ mol^{-1}, whereas $\Delta H^{\ddagger}$ is known to about ± 4 kJ mol^{-1} (assuming k is known to an accuracy of ca. 5 percent at several temperatures). However, $\Delta G^{\ddagger}$ is a strong function of T, whereas $\Delta H^{\ddagger}$ is independent, or nearly independent, of T.

11-2 THE HAMMETT RELATION

Consider first the ionization constant of a series of meta- and para-substituted benzoic acids. For each substituent X a value of σ is computed from the ionization constant of the acid according to the relation

$$\sigma = \log_{10} \frac{K_X}{K_H} \qquad (11\text{-}7)$$

where K_X is the ionization constant for the indicated substituent and K_H is that for the parent compound. This defines the substituent constant for H as $\sigma_H = 0$. Other values[1] are summarized in Table 11-2. Note that groups considered as electron-withdrawing (relative to H) have positive values [for example, $\sigma(p\text{-}CF_3) = +0.54$], whereas those which are electron-releasing are negative [for example, $\sigma(m\text{-}CH_3) = -0.069$].

For any reaction in which such substituted phenyl groups are involved, the rate data (or equilibrium constants, depending on the correlation desired) are considered in light of the following equation:

$$\log \frac{k_X}{k_H} = \rho \sigma_X \qquad (11\text{-}8)$$

where ρ is called the reaction constant. It reflects the sensitivity of the reaction to the variation of substituent; its sign and magnitude are indicative of the electronic substituent effect on the activated complex compared with the reactant. Evaluation of ρ is achieved, as suggested by this relation, through a plot of $\log k_X$ versus σ_X. The latter values are defined by Eq. (11-7); values for a representative set of substituents are given in Table 11-2.

Table 11-2 Hammett substituent constants[a]

X	σ_m	σ_p	X	σ_m	σ_p
NH$_2$	−0.16	−0.66	F	+0.34	+0.06
OH	+0.12	−0.37	Cl	+0.37	+0.23
OCH$_3$	+0.12	−0.27	Br	+0.39	+0.23
SCH$_3$	+0.15	0	I	+0.35	+0.18
CH$_3$	−0.069	−0.17	CF$_3$	+0.42	+0.54
C$_2$H$_5$	−0.07	−0.15	CN	+0.56	+0.66
CH(CH$_3$)$_2$	−	−0.15	NO$_2$	+0.71	+0.78
C(CH$_3$)$_3$	−0.10	−0.20	NMe$_3^+$	+0.88	+0.82
C$_6$H$_5$	+0.06	−0.01	SMe$_2^+$	+1.00	+0.90

[a]Values from Ref. 1.

Hundreds of different reaction series have successfully been correlated by the Hammett equation,[2,3] and each of them is then characterized by a reaction constant ρ.

The qualitative interpretation of ρ is this: A positive value implies a negative charge is developing at the reaction center in the activated complex (cf. the S_N2 reaction of XC$_6$H$_4$CH$_2$Cl with I$^-$, $\rho = +0.79$); the more positive the value the more negative the charge developed at the reaction center (which is the benzylic carbon in this example). Opposite statements apply to a negative value of ρ (cf. the S_N1 reactions of substituted benzyl chlorides with OH$^-$, $\rho = -0.33$).

The reactions of a series of substituted benzenediazonium ions with the bases OH$^-$ and CN$^-$ provide illustrative data,[4] summarized in Table 11-3, for the reactions

$$XC_6H_4N_2^+ + OH^- \longrightarrow syn\text{-}XC_6H_4N_2OH \tag{11-9}$$

$$XC_6H_4N_2^+ + CN^- \longrightarrow syn\text{-}XC_6H_4N_2CN \tag{11-10}$$

Plots for the reactions according to the Hammett equation are shown in Fig. 11-1, which depicts log k_X versus σ. The correlations are seen to be reasonably linear, and

Table 11-3 Rate constants for reactions of aryldiazonium ions with OH$^-$ and CN$^-$,[a] Eqs. (11-9) and (11-10)

Substituent, X	k_{OH}/dm^3 mol^{-1} s^{-1}	k_{CN}/dm^3 mol^{-1} s^{-1}
p-NO$_2$	(5.4 ± 0.3) × 10^5	(1.7 ± 0.2) × 10^4
p-CN	(4.2 ± 0.2) × 10^5	(7.4 ± 0.8) × 10^3
m-CF$_3$	(1.6 ± 0.2) × 10^5	(3.9 ± 0.5) × 10^3
m-Cl	(6.4 ± 0.4) × 10^4	(1.8 ± 0.2) × 10^3
p-Br	(2.1 ± 0.1) × 10^4	(6.5 ± 0.4) × 10^2
p-Cl	(1.6 ± 0.1) × 10^4	(6.8 ± 0.4) × 10^2
H	(4.5 ± 0.5) × 10^3	(2.6 ± 0.2) × 10^2
p-CH$_3$	(1.2 ± 0.2) × 10^3	(9.0 ± 0.9) × 10^1

[a]In aqueous solution at 23°C and ionic strength 0.035 M. Data from Ref. 4.

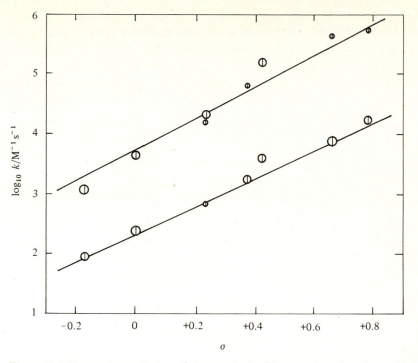

Figure 11-1 Hammett correlation of the reactions of benzenediazonium ions with OH^- [upper line, Eq. (11-9)] and with CN^- [lower line, Eq. (11-10)]. The respective slopes give the values of ρ, which are +2.61 and +2.31.

they lead to these values of ρ: +2.61 for the reaction of OH^- and +2.31 for the reaction of CN^-. The values appear consistent with the nucleophilic process indicated.

Special circumstances such as molecules with direct resonance interactions with the substituents may require a special scale of σ values, as will reactions of saturated compounds. The interested reader is referred to specialized treatments.[3] This author cannot help but feel that the inherent worth of the procedure is its simplicity: One set of substituent constants finds enormously wide applicability. Attempts at modification to accommodate every eventuality are achieved at the expense of additional parameterization. One also recognizes that an attempted Hammett correlation will fail if the mechanism changes at some point along the reaction series.

11-3 THE MARCUS RELATION FOR ELECTRON TRANSFER

The approach is to calculate the rate of an outer-sphere electron transfer reaction[†] from first principles; it is necessary to estimate the contribution to $\Delta G^{\ddagger}$ made by the various steps by which the reaction is thought to occur.[5,6] Those terms include

[†]An outer-sphere reaction mechanism is one in which the reactants do not form an intermediate with bridging functional groups to provide a pathway for electron transfer.

the free energy required to bring reactants to within reactant distance and to reorganize bond distances in each reactant so as to bring each to a common state prior to electron transfer and finally the free-energy change of the net reaction, $\Delta G°$. An abbreviated derivation of the equation is given in an appendix to this chapter.

The popularity of this theory, especially for transition metal complexes, has derived from its use as a means of correlating rates of given reactions with other parameters of interest. For this application we can write the Marcus relation as

$$k_{12} = (k_{11}k_{22}K_{12}f_{12})^{1/2} \qquad (11\text{-}11)$$

where k_{12} and K_{12} are the rate and equilibrium constant for a "cross reaction" such as Eq. (11-12), k_{11} and k_{22} are the self-exchange rate constants for each partner [Eqs. (11-13) and (11-14)], and f_{12} is a factor, usually close to unity, given by Eq. (11-15):

$$Fe(CN)_6^{4-} + IrCl_6^{2-} \xrightarrow{k_{12}} Fe(CN)_6^{3-} + IrCl_6^{3-} \qquad (11\text{-}12)$$

$$Fe(CN)_6^{4-} + {}^*Fe(CN)_6^{3-} \xrightarrow{k_{11}} Fe(CN)_6^{3-} + {}^*Fe(CN)_6^{4-} \qquad (11\text{-}13)$$

$$IrCl_6^{2-} + {}^*IrCl_6^{3-} \xrightarrow{k_{22}} IrCl_6^{3-} + {}^*IrCl_6^{2-} \qquad (11\text{-}14)$$

$$f_{12} = \frac{(\log K_{12})^2}{4 \log (k_{11}k_{22}/Z^2)} \qquad (11\text{-}15)$$

The expression for f contains a quantity Z, the diffusion-controlled rate of collision between uncharged particles in solution, which is often taken as 10^{11} dm^3 mol^{-1} s^{-1}.

Among the uses to which this relation has been put are the following: (1) when the values of k_{11}, k_{22}, K_{12} are known, to affirm that the experimental and calculated values of k_{12} are in accord; compilation of observed and calculated rate constants has been given;[7] (2) to compare different oxidants with the same series of reductants; (3) to check for gross deviations from the theory, which might signal a change in mechanism; (4) to estimate the self-exchange rate constant for one reactant when the value is unknown, based on measured values for one or more cross reactions.[8]

11-4 ACID-BASE CATALYSIS

Many reactions are accelerated by H^+ and/or OH^-. If both are effective catalysts, the apparent rate constant can be expressed as

$$k = k_0 + k_{H^+}[H^+] + k_{OH^-}[OH^-] \qquad (11\text{-}16)$$

The values of the catalytic constants can be evaluated from measurements at pH extremes and checked in intermediate ranges. Relation (11-16) calls for a minimum rate at a hydrogen ion concentration given by

$$[H^+]_{min} = \sqrt{\frac{K_w k_{OH^-}}{k_{H^+}}} \qquad (11\text{-}17)$$

If the rate variation is correctly given by Eq. (11-16), then the reaction is said to show specific acid-base catalysis. An illustration of this phenomenon is provided by kinetic data[9] for the hydrolysis of methyl o-carboxyphenyl acetate (the methyl ester of aspirin):

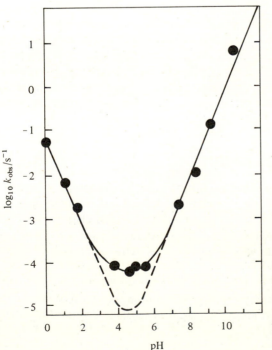

$$\text{(structure with } CO_2CH_3 \text{ and } OCOCH_3) + H_2O = \text{(structure with } CO_2CH_3 \text{ and } OH) + CH_3COOH \qquad (11\text{-}18)$$

A plot of log k_{obs} against pH is shown in Fig. 11-2. The curve shows three portions: (a) a linear portion at pH $\leqslant 2$ with slope -1.00 where the term $k_{H^+}[H^+]$ dominates, (b) a second linear portion at high pH where the term $k_{OH^-}[OH^-]$ dominates, and (c) the intermediate range where the solvent term k_0 or $k_0 a_{H_2O}$ predominates. That the latter term cannot simply be neglected is evident from the dashed line in the figure, which shows the poor fit at intermediate pH values if the solvent term were omitted.

Weak acids and bases are, generally speaking, less effective catalysts than H^+ and OH^-. It is possible to learn whether the buffer components used to maintain a given $[H^+]$ are themselves active catalysts by increasing the concentration of the buffer components HB^+ and B while maintaining the HB^+ to B ratio constant. If an effect is observed, the acid BH^+ and/or the base B are acting as catalysts, and the reaction is referred to as being general acid-base catalyzed.

The correlation of catalytic effectiveness and acid strength (or base strength) can be achieved by a relation similar to the Hammett equation. The traditional form[†] of

[†]These can be seen to have the same general form as Eq. (11-1).

Figure 11-2 The rate constant-pH profile for the hydrolysis of methyl aspirin, Eq. (11-18). The data points are fit to Eq. (11-16) as shown by the solid line, whereas the fit were the solvent path unimportant, $k_0 = 0$, is given by the dashed line. Data from Ref. 9.

the relations, known as the *Brønsted-Pederson equations*, is

$$k_{BH^+} = G_A K_a^{\alpha} \tag{11-19}$$

$$k_B = G_B K_b^{\beta} \tag{11-20}$$

where $0 < \alpha < 1$ and $0 < \beta < 1$. An alternative expression for base catalysis is

$$k_B = \frac{G_B}{K_a^{\beta}} \tag{11-21}$$

These equations require modification if the catalyst contains more than one acidic or basic site. The modified forms are

$$\frac{k_{BH^+}}{p} = G_A \left(\frac{K_a q}{p} \right)^{\alpha} \tag{11-22}$$

$$\frac{k_B}{q} = G_B \left(\frac{K_b p}{q} \right)^{\beta} \tag{11-23}$$

where p is the number of equivalent acidic protons in BH^+ and q is the number of equivalent basic sites in its conjugate base. (For example, oxalic acid and $HC_2O_4^-$, $p = 2$ and $q = 2$; $HC_2O_4^-$ and $C_2O_4^-$, $p = 1$ and $q = 4$; $H_2PO_4^-$ and HPO_4^{2-}, $p = 2$ and $q = 3$.) When p and q might appear ambiguous, the rotational symmetry number provides a better approach.[10]

In practice, one proceeds as follows. Having evaluated k_{BH^+} for a series of acids of similar structure, one plots $\log (k_{BH^+}/p)$ versus $\log (K_a q/p)$. (The "similar structure" proviso is required because acid catalysts of different types—say carboxylic acids, substituted ammonium cations, and metal cations—might each define a different line. The lines are likely to have the same value of the important parameter α but differing values of G_A.) Most often an excellent straight line, whose slope gives the value of α, is obtained. A value outside the range 0.2 to 0.8 can be evaluated only at high buffer concentrations, owing to the term in $[H^+]$. Such determinations require attention to the magnitudes of medium effects. An example of a relatively simple case is found[11] in the rate constant for the acid-catalyzed dehydration of acetaldehyde hydrate,

$$CH_3CH(OH)_2 = CH_3CHO + H_2O \tag{11-24}$$

Values of k_a for both phenols and carboxylic acids are found to be well correlated by Eq. (11-22). Kinetic data,[11] summarized for a representative series of catalysts in Table 11-4, are plotted as $\log (k_a/p)$ versus $(K_a q/p)$ in Fig. 11-3. The slope of the line provides a value of α, which in this case is equal to 0.54. It should be noted that the rate constants for certain other acid catalysts of different structural classes do not lie in that line. Presumably, specific interactions, such as different degrees of resonance stabilization which become important when structure type is changed, account for the divergences.

The chemical interpretation of α is that it reflects the sensitivity of rate to the acid strength of the catalyst and thereby the degree of proton transfer to substrate in the activated complex. A value of α above 0.8 signifies substantial proton transfer, and vice versa.

Table 11-4 Rate constants[a] for the acid-catalyzed dehydration of acetaldehyde hydrate

Acid catalyst (p,q)	$k_a/M^{-1}\ s^{-1}$	K_a	$\log_{10}(k_a/p)$	$\log_{10}(K_a q/p)$
Thymol (1,1)	1.87×10^{-4}	3.2×10^{-11}	-3.73	-10.49
p-Dihydroxybenzene (2,1)	2.17×10^{-4}	4.5×10^{-11}	-3.96	-10.64
Phenol(hydroxybenzene (1,1)	3.02×10^{-4}	1.06×10^{-10}	-3.52	-9.97
m-Dihydroxybenzene (2,1)	4.33×10^{-4}	1.55×10^{-10}	-3.66	-10.11
o-Dihydroxybenzene (2,1)	1.02×10^{-4}	6.6×10^{-10}	-3.29	-9.48
o-Nitrophenol (1,1)	5.57×10^{-3}	6.8×10^{-8}	-2.25	-7.17
2,4,6-Trichlorophenol (1,1)	2.55×10^{-2}	3.9×10^{-7}	-1.59	-6.41
Propionic acid (1,2)	3.00×10^{-1}	1.35×10^{-5}	-0.523	-4.57
o-Methylbenzoic acid (1,2)	6.4×10^{-1}	1.22×10^{-4}	-0.195	-3.61
Formic acid (1,2)	7.25×10^{-1}	1.77×10^{-4}	-0.140	-3.45
Benzoic acid (1,2)	9.9×10^{-1}	6.6×10^{-5}	-0.006	-3.88
Phenoxyacetic acid (1,2)	1.53×10^{0}	7.6×10^{-4}	0.186	-2.82
Chloroacetic acid (1,2)	2.43×10^{0}	1.51×10^{-3}	0.386	-2.52
Dichloroacetic acid (1,2)	1.29×10^{1}	5.0×10^{-2}	1.11	-1.00

[a]Rate constants k_a (from Ref. 11), determined in acetone-water at 25.0°C; acid ionization constants in water at 25°C.

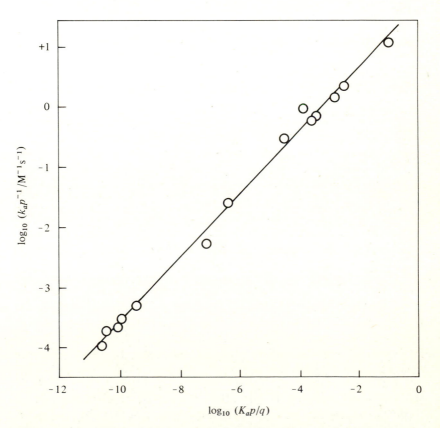

Figure 11-3 Correlation of the rate constants for the acid-catalyzed dehydration of acetaldehyde hydrate by the Brønsted-Pedersen catalysis law. Data from Table 11-4 and Ref. 11.

One mechanistic scheme to be considered is the following:

$$S + BH^+ \underset{-1}{\overset{1}{\rightleftharpoons}} SH^+ + B \tag{11-25}$$

$$SH^+ \xrightarrow{2} P + H^+ \tag{11-26}$$

The rate law, making the steady-state approximation for $[SH^+]$, is

$$\frac{d[P]}{dt} = \frac{k_1[S][BH^+]}{1 + k_{-1}[B]/k_2} \tag{11-27}$$

Depending on the relative values of $k_{-1}[B]$ and k_2, either of two limiting forms may be found, $k_1[S][BH^+]$ and $(k_1 k_2/k_{-1})([S][BH^+]/[B])$. The latter is equivalent to $(k_1 k_2/K_a k_{-1})([S][H^+])$, where K_a is the ionization constant of BH^+. This example illustrates one problem in making the distinction between general and specific acid catalysis.

The reader, who will recognize the abbreviated nature of the treatment here, is directed to more complete accounts of the subject.[12,13]

11-5 STEREOCHEMISTRY AND THE ACTIVATION PROCESS

The direction of approach of two reactants along the reaction coordinate is, except for the simplest of reactants, generally an unknown quantity. Although sometimes settled rather simply on the basis of what groups are bonded to one another in the products, that is not a general solution. For groups of any complexity at all, the question of the geometry of the activated complex can best be resolved by use of chiral reagents.

The example familiar to everyone is the Walden inversion at saturated carbon accompanying bimolecular nucleophilic substitution. Because it and other simple organic reactions are familiar, the examples chosen for mention here consist of less familiar systems.

(a) The rate of racemization of the optically active complex D-cis-Co(en)$_2$Cl$_2^+$ is equal to the rate of exchange of one bound chloride with free radiochloride ions,[14] implicating the same activated complex from which is derived an intermediate that is necessarily symmetrical.

(b) The reaction of the erythro diastereomer of $(\eta\text{-}C_5H_5)Fe(CO)_2R^\dagger$ with iodine in carbon disulfide converts the complex to >90% threo, whereas reaction with chlorine in chloroform gives >95 percent retention:[15]

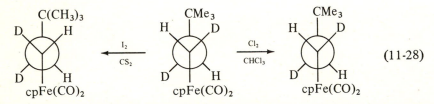

$$\tag{11-28}$$

†The group R is Bu^{t}CHDCHD-, a bulky group, such as t-butyl or sometimes phenyl, being used to favor the gauche conformation and thus lead to a ready interpretation of the nmr spectrum.

This example is useful in another context as well, because it introduces a method other than optical activity for assessing the stereochemical course of a reaction. The coupling constant between the vicinal hydrogens is greatly different[16] in erythro and threo diastereomers, permitting a distinction based on nmr methods.

(c) Electron transfer can catalyze racemization. Traces of the substitution-labile Co(II) complex lead to rapid loss of optical activity of D-Co(en)$_3^{3+}$, undoubtedly via the process[17]

$$D\text{-}Co(en)_3{}^{3+} + L\text{-}Co(en)_3{}^{2+} \rightleftharpoons D\text{-}Co(en)_3{}^{2+} + L\text{-}Co(en)_3{}^{3+}$$

$$\downarrow \text{Fast} \tag{11-29}$$

$$\pm\text{-}Co(en)_3{}^{2+}$$

11-6 ISOTOPE RATE EFFECTS

Isotopes find a number of uses in mechanistic studies, including innumerable instances in which labeling experiments have permitted a greater insight into mechanism. This section concerns the difference in reaction rate between one compound and another that is identical except for an isotopic label.

The major factor in determining the magnitude of the kinetic isotope effect lies in the difference in the zero-point energy of the two bonds. Owing to the mass differences, the rate constants will be dependent on the isotopic masses, the more so (1) the greater the bond in question is involved in the activation process and (2) the isotopic substitution involves lighter atoms.

A rigorous treatment of the phenomenon based on activated complex theory is possible although complex.[18,19] We shall content ourselves with a highly approximate treatment. Consider a reaction in which the main activation process is the stretching of a single bond. Because that is the reaction coordinate, the transition states have the same energy for both isotopes and the rate difference lies totally in the two ground states. The ratio of stretching frequencies is given by the harmonic oscillator equation as a function of the reduced masses of the hypothetical diatomic molecule derived from the two atoms making up the bond:

$$\frac{v_1}{v_2} \cong \left(\frac{\mu_2}{\mu_1}\right)^{1/2} \tag{11-30}$$

When a new bond is formed as the old is broken a slightly different treatment is required. In this case the ratio is roughly the square root of the ratio of masses of the two isotopically different atoms alone. For proton transfer, the calculated frequency ratio will approach $2^{1/2}$.

The frequency ratios can be converted into rate constant ratios combining this result with the expression for the zero-point energy, $E = hv/2$, and the Arrhenius equation. Consider a CH bond, for which we know a typical stretching frequency is 2900 cm^{-1}. That converts to $v_H = 8.70 \times 10^{13}$ s^{-1}. We take $v_D = v_H/2^{1/2} = 6.15 \times 10^{13}$ s^{-1}. Consequently, the difference of zero-point energies is $E_H - E_D = 8.4 \times 10^{-21}$ J molecule^{-1} or 5.1×10^3 J mol^{-1}. The expected value of k_H/k_D at 298 K is

$$\frac{k_H}{k_D} \simeq \exp\left(-\frac{E_H - E_D}{RT}\right) = 7.8 \tag{11-31}$$

Inclusion of bending motions and use of more exact vibrational expressions can alter this approximate result. Values as high as $k_H/k_D \sim 20$ are possible, although unlikely.

APPENDIX: A DERIVATION OF THE MARCUS EQUATION FOR ELECTRON TRANSFER

The basis of the Marcus relation is the following. The overall process is depicted in Fig. 11-4, which shows the different contributions to the free energy of activation for forward and reverse reactions. The terms are the following: (1) the work required to bring the reactants from their ground states to their reaction distance (ω_r) and then to remove the newly formed products into the bulk solvent ($-\omega_p$), (2) the free energy of the process by which the inner and outer coordination spheres are converted to the orientation which prevails in the transition states (λ_r) and a similar contribution by which the newly formed products relax to their equilibrium configurations ($-\lambda_p$), (3) the actual electron transfer step, for which no further free energy of activation is required, since reactants have been brought to the same energy as the activated complex.

It is evident from these considerations or from Fig. 11-4 that the following rela-

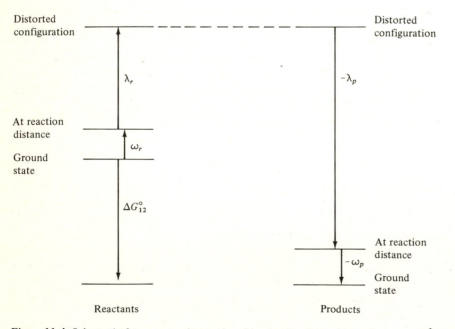

Reactants Products

Figure 11-4 Schematic free energy changes in a bimolecular electron transfer reaction.[6] The various contributions are discussed in the text.

tions hold:

$$\Delta G_{12}^{\circ} = \Delta G_{12}^{*} - \Delta G_{21}^{*} \qquad (11\text{-}A1)$$

$$\Delta G_{12}^{\circ} = \omega_r - \omega_p + \lambda_r - \lambda_p \qquad (11\text{-}A2)$$

$$\Delta G_{12}^{*} = \omega_r + \lambda_r \qquad \Delta G_{21}^{*} = \omega_p + \lambda_p \qquad (11\text{-}A3)$$

The energy required to reach a configuration where the electron can be transferred without additional activation, ΔG_{12}^{*}, is closely related to the conventional free energy of activation

$$\Delta G_{12}^{*} = \Delta G_{12}^{\ddagger} - RT \ln \frac{NhZ}{RT} \qquad (11\text{-}A4)$$

and thus to the second-order rate constant k

$$k_{12} = Z \exp \left(\frac{-\Delta G_{12}^{*}}{RT} \right) \qquad (11\text{-}A5)$$

where Z represents the diffusion-controlled rate of collision between uncharged particles in solution and is often taken as 10^{11} dm^3 mol^{-1} s^{-1}.

Following the simplified derivation of Newton,[6] it is first noted that the electrostatic work terms ω_r, etc., are generally small, particularly in ionic solutions; moreover, the values of ω_r and ω_p, being similar, tend to offset one another. That is particularly true when, as will soon be developed, one does not attempt to calculate $\Delta G_{12}^{\ddagger}$ (or k_{12}) from first principles, but only in reference to the values for the two self-exchange processes, $\Delta G_{11}^{\ddagger}$ and $\Delta G_{22}^{\ddagger}$. On that basis we have

$$\Delta G_{12}^{\circ} \approx \lambda_r - \lambda_p \qquad \Delta G_{12}^{*} \approx \lambda_r \qquad \Delta G_{21}^{*} \approx \lambda_p \qquad (11\text{-}A6)$$

The general equation for the electron transfer process can be written as

$$M_1 + N_2 = M_2 + N_1 \qquad (11\text{-}A7)$$

where the subscript 1 represents the reduced forms of chemical substances M and N and subscript 2 the oxidized form. Attention is directed to those distortions of the coordination shells of each substance which occur along the reaction coordinate. These are likely to resemble vibrational "breathing" modes, and it is assumed here that they can be represented by a single parameter d. Four potential functions, d_{M_1}, d_{M_2}, d_{N_1}, and d_{N_2}, must thus be considered. Since such functions are not well known, an approximation is made that they are parabolic and that the force constants do not depend on the oxidation state of M or N.[†] If the units of length are normalized such that d values are unity for the undistorted configurations,

$$\begin{aligned}
\lambda_{M_1} &= \tfrac{1}{2} f_M (x-1)^2 \\
\lambda_{M_2} &= \tfrac{1}{2} f_M x^2 \\
\lambda_{N_2} &= \tfrac{1}{2} f_N y^2 \\
\lambda_{N_1} &= \tfrac{1}{2} f_N (y-1)^2
\end{aligned} \qquad (11\text{-}A8)$$

[†]The more rigorous derivation[5] does not require this approximation, nor does it require that reorganization energy be expressed in terms of a single parameter d.

substitution into Eq. (11-A6) gives

$$\Delta G_{12}^{\circ} \approx \tfrac{1}{2} f_M - f_M x - \tfrac{1}{2} f_N + f_N y \tag{11-A9}$$

If distortions about M and N are independent of one another, a parameter α may be defined such that

$$\alpha \Delta G_{12}^{\circ} \approx \tfrac{1}{2} f_M - f_N x$$
$$(1 - \alpha)\Delta G_{12}^{\circ} \approx -\tfrac{1}{2} f_N + f_N y \tag{11-A10}$$

Substitution into Eq. (11-A6) gives

$$\Delta G_{12}^{*} = \tfrac{1}{2} f_M (x - 1)^2 + \tfrac{1}{2} f_N y^2 \tag{11-A11}$$

$$\Delta G_{12}^{*} = \frac{f_M + f_N}{8} + \frac{\Delta G_{12}^{\circ}}{2} + \left[\frac{\alpha^2}{2 f_M} + \frac{(1 - x)^2}{2 f_N} \right] (\Delta G_{12}^{\circ})^2 \tag{11-A12}$$

The solution of interest is that for which the free energy of activation is a minimum, or $\partial \Delta G_{12}^{*}/\partial \alpha = 0$; differentiation of Eq. (11-A12) and solution for α gives

$$\alpha = \frac{f_M}{f_M + f_N} \tag{11-A13}$$

and substitution of this expression into (11-A12) yields the final result

$$\Delta G_{12}^{*} = \frac{f_M + f_N}{8} + \frac{\Delta G_{12}^{\circ}}{2} + \frac{(\Delta G_{12}^{\circ})^2}{2(f_M + f_N)} \tag{11-A14}$$

The immediate goal is to relate the activation free energy for the "cross reaction" ΔG_{12}^{*} to that for similarly expressed values relating to the self-exchange reactions. For the latter, $f_M = f_N$ and $\Delta G_{11}^{\circ} = \Delta G_{22}^{\circ} = 0$, and Eq. (11-A14) becomes

$$\Delta G_{11}^{*} = \frac{f_M}{4} \quad \text{and} \quad \Delta G_{22}^{*} = \frac{f_N}{4} \tag{11-A15}$$

In terms of these comparative parameters, the expression for ΔG_{12}^{*} becomes

$$\Delta G_{12}^{*} = \frac{1}{2} \left\{ \Delta G_{11}^{*} + \Delta G_{22}^{*} + \Delta G_{12}^{\circ} + \frac{(\Delta G_{12}^{\circ})^2}{4[(\Delta G_{11}^{*})^2 + (\Delta G_{22}^{*})^2]} \right\} \tag{11-A16}$$

After substitutions from Eqs. (11-15), (11-A4), and (11-A5), there results the usual form in which the Marcus equation is expressed,

$$k_{12} = (k_{11} k_{22} K_{12} f_{12})^{1/2} \tag{11-A17}$$

REFERENCES

1. Brown, H. C., and Y. Okamoto: *J. Am. Chem. Soc.*, **80**:4979 (1958).
2. Jaffé, H. H.: *Chem. Revs.*, **53**:191 (1953).
3. Wells, P. R.: "Linear Free Energy Relationships," Academic Press, Inc., London, 1968.
4. Ritchie, C. D., and D. J. Wright: *J. Am. Chem. Soc.*, **93**:6574 (1971).

5. Marcus, R. A.: *Ann. Rev. Phys. Chem.*, **15**:155 (1964); *J. Phys. Chem.*, **72**:891 (1968).
6. Newton, T. W.: *J. Chem. Educ.*, **45**:571 (1968).
7. Pennington, D. E.: in A. E. Martell (ed.), "Coordination Chemistry," vol. 2, monograph 174, American Chemical Society, Washington, D.C., 1978, pp. 482–483.
8. Pladziewicz, J. R., and J. H. Espenson: *J. Am. Chem. Soc.*, **95**:56 (1973).
9. Pierre, T. S., and W. P. Jencks: *J. Am. Chem. Soc.*, **90**:3817 (1968).
10. Benson, S. W.: *J. Am. Chem. Soc.*, **80**:5151 (1958).
11. Bell, R. P., and W. C. E. Higginson: *Proc. Roy. Soc. London*, **A197**:141 (1949).
12. Bender, M. L.: "Mechanisms of Homogeneous Catalysis from Protons to Proteins," John Wiley & Sons, Inc., New York, 1971.
13. Bell, R. P.: "The Proton in Chemistry," 2d ed., Cornell University Press, Ithaca, N.Y., 1973.
14. Brown, D. D., and R. S. Nyholm: *J. Chem. Soc.*, 2696 (1953).
15. Whitesides, G. M., and D. J. Boschetto: *J. Am. Chem. Soc.*, **91**:4313 (1969); **93**:1529 (1971).
16. Bock, P. L., D. J. Boschetto, J. R. Rasmussen, J. P. Demers, and G. M. Whitesides: *J. Am. Chem. Soc.*, **96**:2814 (1974).
17. Busch, D. H.: *J. Am. Chem. Soc.*, **77**:2747 (1955).
18. Melander, L.: "Isotope Effects on Reaction Rates," The Ronald Press Company, New York, 1960.
19. Bigeleisen, J., and M. Wolfsberg: *Advances in Chem. Phys.*, vol. I, pp. 15–76, 1958.

PROBLEMS

11-1 *Hammett correlation.* Carry out a Hammett correlation for the reaction of substituted benzoic acids with diazophenylamine. Values from Benkeser et al., *J. Am. Chem. Soc.*, **78**:682 (1956), in ethanol at 303 K are as follows:

Substituent	m-CH$_3$	H	m-OCH$_3$	p-Cl	p-Br	m-Br	m-NO$_2$	p-NO$_2$
$10^2\,k/\text{dm}^3\,\text{mol}^{-1}\,\text{s}^{-1}$	1.60	1.77	2.14	2.77	3.30	4.31	8.62	8.80

11-2 *Hammett equation.* What form does activated complex theory give for the temperature dependence of Hammett's ρ for a series in which $\Delta S^{\ddagger}$ is essentially constant, as is often the case?

11-3 *Acid catalysis.* The acid-catalyzed bromination of an ester RCH_2COOR' has $\alpha = 0.5$. The monosuccinate acid ion, $K_a = 2.4 \times 10^{-6}$, has $k = 3.6 \times 10^4$ M^{-1} s^{-1}. Estimate k for formic acid, $K_a = 2.1 \times 10^{-4}$.

11-4 *Base catalysis.* Compute G_B and β from the following data for the base-catalyzed enolization of 3-methylacetone [D. B. Dahlberg and F. A. Long, *J. Am. Chem. Soc.*, **95**:3825 (1973)]:

Catalyst	ClCH$_2$CO$_2^-$	CH$_3$CO$_2^-$	HPO$_4^{2-}$
K_B	7.2×10^{-12}	5.7×10^{-10}	1.6×10^{-7}
$k/\text{dm}^3\,\text{mol}^{-1}\,\text{s}^{-1}$	1.41×10^{-3}	1.34×10^{-2}	2.60×10^{-1}

11-5 *Substituent effects.* Many series of related reactions are characterized by a linear relation between $\Delta H^{\ddagger}$ and $\Delta S^{\ddagger}$:

$$\Delta H^{\ddagger} = \beta\,\Delta S^{\ddagger} + C$$

What is the significance of this relation, and in particular what is the meaning of the slope β?

11-6 *Hammett correlation.* Berliner and Monack [*J. Am. Chem. Soc.*, **74**:1574 (1952)] have evaluated rate constants for reactions of a group of para-substituted 2-nitrophenyl bromides with the base piperidine (B). Use their data to estimate the reaction constant ρ and σ for CO_2^-

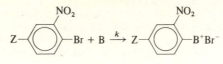

Substituent Z	Br	CO_2^-	H	CH_3	NH_2
k at 298 K/s^{-1}	3.8×10^{-4}	1.2×10^{-4}	4.8×10^{-5}	7.0×10^{-6}	6.0×10^{-9}

11-7 *LFER.* Consider the rate of base hydrolysis of a series of ethyl benzoates given by $d[C_2H_5OH]/dt = k_X[XC_6H_4COOC_2H_5][OH^-]$. Show that a plot of $\log k_X$ versus $\log K_a$, where K_a is the ionization constant of the parent benzoic acid, should be linear, and relate its slope to that given by a conventional Hammett plot of $\log k$ versus σ.

11-8 *Specific and general acid catalysis.* Formulate the rate law for the acid-catalyzed hydrolysis reaction

$$AcNHC(CH_3)=CH_2 + H_2O = HOAc + H_2NC(CH_3)=CH_2$$

and evaluate the rate constant(s) from the following results in HOAc-NaOAc buffers given by J. M. Csizmadia, *J. Am. Chem. Soc.*, **101**:883 (1979).

pH	10^2 [HOAc]/M	10^2 [OAc$^-$]/M	10^3 k_{obs}/s^{-1}
3.93	5.00	1.00	1.86
4.04	4.00	1.00	1.53
4.20	3.00	1.00	1.17
4.65	0.250	0.250	0.275
4.65	0.500	0.500	0.313
4.65	1.00	1.00	0.390
4.65	2.00	2.00	0.530
4.65	3.00	3.00	0.670
4.95	0.500	1.00	0.194

ANSWERS TO SELECTED NUMERICAL PROBLEMS

Chapter 2

2-2. (a) 1.25×10^{-2} s^{-1} (a slight function of τ); (b) 0.014 to 0.030 (depending on τ).

2-3. 121 min

2-5. 1.57×10^3 dm^3 mol^{-1} s^{-1}

2-6. 6.1×10^{-6} s^{-1}

2-7. $k_{23} - k_{34} = 89$ dm^3 mol^{-1} s^{-1}

2-8. $k = \{15.4$ M^{-1} s$^{-1} + (2.4 \times 10^2$ M^{-2} s$^{-1})[N_2H_4]\} [N_2H_4]$

2-10. 16.8 dm^3 mol^{-1} s^{-1}

Chapter 3

3-1. $k_f = 5.56 \times 10^{-3}$ dm^3 mol^{-1} s^{-1}; $k_r = 1.25 \times 10^{-3}$ dm^3 mol^{-1} s^{-1}

3-3. $k_1 = 0.173$ dm^3 mol^{-1} s^{-1}; $k_{-1} = 1.17 \times 10^{-2}$ s^{-1}; $k_{obs} = 2.2 \times 10^{-2}$ s^{-1}

3-6. (a) 88 s; (b) 1.3×10^{-5} mol dm^{-3} s^{-1}; (c) 1.05 dm^3 mol^{-1} s^{-1}; (d) 66 s

3-7. (a) 9.7×10^{-5} mol dm^{-3} s^{-1}; (b) 1.32×10^{-2} dm^3 mol^{-1} s^{-1}; (c) 306 s

3-8. 2.9×10^{-3} dm^3 mol^{-1} s^{-1}

3-10. $k_1/k_2 = 0.59$ mol dm^{-3}

3-11. (b) $k_1 = 1.7 \times 10^{-4}$ dm^3 mol^{-1} s^{-1}; $k_2 = 1.1 \times 10^{-5}$ s^{-1}

3-12. (b) 1.1×10^{-4} s^{-1}, 4.3×10^{-4} s^{-1}; (c) 0.0370, 0.0210

3-13. $k_1 = 10.7$ s^{-1}; $k_{-1} = 8.7 \times 10^2$ dm^3 mol^{-1} s^{-1}

Chapter 4

4-6. 9×10^3 dm^3 mol^{-1}

4-7. (c) $k_1 = 7.3 \times 10^{-4}$ s^{-1}, $k_{-1}/k_S = 2.31$; $k_N/k_S = 8.22$ dm^3 mol^{-1}

4-8. (c) $k_2/k_1 = 1.54$; $k_3/k_{-1} = 2.64$

4-9. (c) (1) 0.018; (2) 0.500; (3) 12.4 dm^3 mol^{-1}; (4) 1.06×10^{-3} s^{-1}; (5) 8.6

Chapter 5

5-2. (b) (1), $-E_1/R$; (2), $-(E_{-1} + E_2)/R$
5-6. $k_1 = 5.5 \times 10^{-3} \text{ s}^{-1}$; $k_{-1}/k_2 = 65$
5-16. $k[U(VI)][U^{4+}][H^+]^{-3}$, $k = 2.1 \times 10^{-7} \text{ mol}^2 \text{ dm}^{-6} \text{ s}^{-1}$

Chapter 6

6-1. (a) 3.79; (b) $k_f = 6.65 \times 10^{-3}$, $k_r = 1.75 \times 10^{-3}$; (c) $\Delta H_f^{\ddagger} = 11.2 \text{ kJ mol}^{-1}$, $\Delta S_f^{\ddagger} = -250 \text{ J mol}^{-1} \text{ K}^{-1}$; $\Delta H_r^{\ddagger} = 23.0 \text{ kJ mol}^{-1}$; $\Delta S_r^{\ddagger} = -223 \text{ J mol}^{-1} \text{ K}^{-1}$; (d) $\Delta H^{\circ} = -11.9 \text{ kJ mol}^{-1}$; $\Delta S^{\circ} = -27 \text{ J mol}^{-1} \text{ K}^{-1}$
6-2. $k_r[UO_2^+]^2[H^+]$, $k_r = 1.05 \times 10^3 \text{ dm}^6 \text{ mol}^{-2} \text{ s}^{-1}$
6-3. (a) 0.71 $\text{dm}^3 \text{ mol}^{-1} \text{ s}^{-1}$
6-4. $k_2'/k_1' = 1.286$; $\Delta H^{\ddagger} = 83.81 \text{ kJ mol}^{-1}$
6-6. (a) $3.1 \times 10^{-5} \text{ s}^{-1}$; (b) $1.02 \times 10^{-4} \text{ mol dm}^{-3}$; (c) $2.2 \times 10^{-5} \text{ s}^{-1}$
6-8. $1.5 \times 10^{12} \text{ s}^{-1}$
6-9. 1.8×10^{-3} (Ar), 2.4×10^{-3} (He), 2.7×10^{-3} (N_2); 3.1×10^{-3} (O_2); 6.7×10^{-3} (CO_2); 7.1×10^{-3} (O_3)

Chapter 7

7-1. (b) $\log(k_2/\text{atm}^{-1} \text{ s}^{-1}) = 9.38 - 26{,}500 \text{ cal mol}^{-1}/RT$
7-10. (a) $3.0 \times 10^{-17} \text{ mol dm}^{-3}$; (b) $5.7 \times 10^{-19} \text{ mol dm}^{-3}$

Chapter 8

8-1. (a) $5.2 \times 10^{28} \text{ cm}^{-3} \text{ s}^{-1}$; (b) $5.2 \times 10^{15} \text{ molecules cm}^{-3}$
8-2. (a) $\sim 10^{-4}$; (b) 1.8 nm
8-5. (b) $1.15 \times 10^{-4} \text{ s}^{-1}$
8-6. (a) $1.4 \times 10^{14} \text{ cm}^3 \text{ mol}^{-1} \text{ s}^{-1}$; (b) $2.5 \times 10^{13} \text{ s}^{-1}$; (c) 0.18 mol cm^{-3}; (d) $3.5 \times 10^2 \text{ cm}^3 \text{ mol}^{-1} \text{ s}^{-1}$; (e) 5
8-8. $p_{1/2} = 8 \times 10^{10} \text{ torr}$
8-9. (b) $1.6 \times 10^{14} \text{ cm}^3 \text{ mol}^{-1} \text{ s}^{-1}$; (c) $1.6 \times 10^{12} \text{ cm}^3 \text{ mol}^{-1} \text{ s}^{-1}$

Chapter 9

9-2. 1+
9-3. $k_1 = 1.36 \times 10^{-5} \text{ dm}^3 \text{ mol}^{-1} \text{ s}^{-1}$, $k_2 = 9.2 \times 10^{-8} \text{ s}^{-1}$
9-5. $4.1 \pm 0.5 \text{ cm}^3 \text{ mol}^{-1}$
9-6. $5.6 \text{ cm}^3 \text{ mol}^{-1}$

Chapter 10

10-4. $k_f = 9.1 \times 10^7 \text{ dm}^3 \text{ mol}^{-1} \text{ s}^{-1}$; $k_r = 4.5 \times 10^4 \text{ s}^{-1}$

Chapter 11

11-1. $\rho = 0.92$

11-3. $2.4 \times 10^{-3} \text{ dm}^3 \text{ mol}^{-1} \text{ s}^{-1}$

11-4. $G_B = 3 \times 10^2; \beta = 0.50$

11-6. $\rho = +4.2; \sigma(CO_2^-) = +0.12$

11-8. $k/\text{s}^{-1} = 10.7 \text{ [H}^+] + 0.014 \text{ [HOAc]}$

INDEX

Absolute rate theory (*see* Activated complex theory)

Acid-base catalysis:
  Brønsted-Pedersen relations, 194
  general, 200–201
  mechanism for, 199–202
  specific, 198–199

Activated complex:
  elemental composition of, 90–91
  versus intermediate, 90

Activated complex theory:
  applied to bimolecular reactions, 155–157
  applied to reactions in solutions, 170
  calculations using, 157–159
  $\Delta S^{\ddagger}$ from, 157–159
  derivation of, 153–155
  Eyring equation for, 117–118
  and temperature effects, 117–120

Activation energy, 117, 139
  apparent, 121–122
  Arrhenius, 117
  and $\Delta H^{\ddagger}$, 120

Activation parameters, 8
  accuracy of, 195
  apparent, for complex mechanisms, 121–122
  mechanism and, 122
  from pressure dependence, 177–179
  from temperature dependence, 117–120
  variation with temperature, 125–128

Ad eosdem competition, 59

Ad eundem competition, 58–59

Alternative mechanisms:
  electrical analogs for, 100–101
  equivalent expressions for, 94–95
  formulated from rate law, 93
  having consecutive reactions, 99–101
  salt effects and, 176–177
  tests for, 105–106

Arrhenius activation energy, 117

Arrhenius equation, 117–119
  applied to composite rate constants, 121–122

Arrhenius plots, 119
  for complex mechanisms, 121–122
  curvature in, 121–122, 125–128

Biphasic reaction, 65–72
  dual solutions for, 69–72
  resembling a monophasic, 70–72

Branching chain reactions, 142–143

Brønsted-Bjerrum equation, 170

Brønsted-Pedersen relations, 194

Bunnett methods, 58–59
  (*See also* Acid-base catalysis)

Catalysis:
  by acids and bases, 194–202
  by enzymes, 80–82

Catalysis *(Cont.):*
  kinetic data for, 81–82
  on surfaces, 82–83
Chain length:
  of acetaldehyde decomposition, 137
  defined, 136
Chain mechanisms, 135
Chain-propagating steps, 135
Chain reactions, 134–143
  and activation energies, 137–139
  branching, 142–143
  free radicals in, 140
  initiation of, 134
  photochemically initiated, 139
  propagation of, 135
  steady-state approximations for, 135, 139
  termination of, 135
Coalescence temperature, 187–188
Collision theory:
  for bimolecular reactions, 150–152
  steric factors for, 152
  and temperature effects, 117
Competition methods:
  ad eosdem, 58–59
  ad eundem, 57
  for intermediates in concurrent reactions,
    58–59
Competitive inhibition, 82
Complex, activated, 90–91
Concentration-jump method, 48–50
Concentration units, 3
  for rate equations, 120
  and value of $\Delta S^\ddagger$, 179–180
Concurrent reactions, 51–59
  competition methods for, 58–59
  of mixtures, 57
  product yields in, 57–59
Consecutive reactions:
  biphasic, 65–72
  first-order, 65–72
  induction period, 66, 69
  resembling a single step, 70–72
  revealed in rate law, 92–93
  secular equilibrium, 66
Constant:
  composite rate, 121–122
  dielectric, 167, 170–172
  maximum rate, 121–122
  Michaelis, 81
  rate, 14–15, 43
  reaction, 195–196
Constant ionic strength principle, 174–
  175

Debye-Hückel equation:
  alternative forms of, 172–173
  and salt effects, 172, 175
$\Delta C_p^\ddagger$ *(see* Heat capacity of activation)
$\Delta G^\ddagger$ *(see* Free energy of activation)
$\Delta H^\ddagger$ *(see* Enthalpy of activation)
$\Delta S^\ddagger$ *(see* Entropy of activation)
$\Delta V^\ddagger$ *(see* Volume of activation)
$(\Delta z^2)^\ddagger$ (ionic charge of activation), 175
Dielectric constant:
  effect illustrated, 167
  and reaction rates, 170–172
Diffusion-controlled reactions:
  equations for rates limited by, 168–169
  influence of ionic charge on, 169
  and maximum rate constant, 96–97

E_a *(see* Activation energy)
Electron transfer, Marcus equation for, 197–
  198, 204–206
Elementary reaction, 5
"Encounter" in solution reactions, 167
End-point reading *(see P_∞)*
Energy, free:
  of activation, 125, 195
  of reaction, 125
Enthalpy of activation, $(\Delta H^\ddagger)$, 117–120
  accuracy, compared to $\Delta G^\ddagger$, 195
  for composite rate constants, 121–122
  differential temperature methods, 132
  and E_a, 120
  in parallel-path mechanisms, 121–122
  precise value for, 126–128, 132
  and reaction mechanism, 122
  variation with temperature, 125–128
Entropy of activation $(\Delta S^\ddagger)$, 117–120
  from activated complex theory, 157–159
  and concentration units, 179–180
  and reaction mechanism, 122–123
  standard state and, 120, 179–180
  theoretical calculation of, 157–159
Enzyme-catalyzed reactions, 80–82
Equation(s):
  Arrhenius, 117–119, 121–122
  Brønsted-Bjerrum, 170
  Debye-Hückel, 172–173, 175
  Eyring, 117–118
  Fick's diffusion, 168
  Hammett, 195–197
  McKay, 51–54
  Marcus, 197–198, 204–206
  Taft, 194

Exchange reactions, 51–55
  examples of, 51
  half-times for, 53
  McKay equation for, 51–54
  mechanisms for, 52–53
  and microscopic reversibility, 129–131
  by quenched-flow technique, 184
Extent of reaction:
  of exchanges, 52
  and reaction order, 34–35
Eyring equation (*see* Activated complex theory)

Fast reactions:
  techniques for, 182–183
  time resolution of, 183
Fick's diffusion equations, 168
First-order kinetics, 12–15
  estimation of P_∞, 24–28
  evaluation of rate constants, 14–15
  Guggenheim method, 25–28
  irreversible, 12–15
  Kezdy-Swinbourne method, 25–28
  using instrumental methods, 22–25
Flash photolysis:
  chemical transients produced by, 188–189
  procedure for, 188
Flooding, technique of:
  excess reagent, 30–32
  and impurities, 32
  in pseudo-first-order reactions, 12–15
  in reversible reactions, 45, 47–48
Flow methods:
  for rapid reactions, 183–184
  stopped-flow, 184
Fractional-time method, 34
Free energy of activation ($\Delta G^\ddagger$):
  accuracy, compared to $\Delta H^\ddagger$, 195
  calculation of, 125
  and free energy of reaction, (ΔG°), 125
  relation to $\Delta H^\ddagger$ and $\Delta S^\ddagger$, 125
Free radicals:
  in chain reaction, 140
  from pulse radiolysis, 189–190
  scavenging of, 57–58, 189–190
Frequency factor, 117

Guggenheim method, 125–128

Half-life (*see* Half-time)

Half-time, 14, 24
  and reaction order, 24, 34
  of reversible reactions, 43–44
Hammett equation:
  correlations in terms of, 195–197
  illustrated, 197
  substituent constants for, 195–196
Hanes plot, 88
Heat capacity of activation (ΔC_p), 122
  calculation of, 126–128
  defined, 125
  magnitude of, 125
Heterogeneous reactions, 82–83

Induction period, 66, 69
Initial rates, 7, 81
Initiation:
  of chain reactions, 134
  photochemical, 139
Integrated rate laws, 30, 35–37
Intermediate (*see* Reaction intermediate)
Inverse reaction orders and mechanisms, 92–93
Ionic charge of activation $(\Delta z^2)^\ddagger$, 175
Ionic strength:
  defined, 172
  and ionic reactions, 172–176
  principle of constant, 174–175
Ions, reactions of:
  dielectric constant effect on, 171–172
  salt effects on, 172–176
Isotopes, 203–204
Isotopic exchange (*see* Exchange reactions)

Kinetic isotope effect, 203–204
Kinetics:
  first-order, 12–15, 22–28
  second-order, 16–21, 24, 27–29

Langmuir adsorption isotherm, 82–83
LFER (linear free-energy relation):
  applications of, 195
  by Hammett equation, 195–197
  by Marcus relation, 197–198, 204–206
  methods for, tabulated, 194
Lindemann mechanism:
  modified by RRKM, 162
  for unimolecular reaction, 160
Linear free-energy relation (*see* LFER)
Lineweaver-Burk plot, 88
Long-chain approximation, 137

McKay equation, 51–54
Magnetic resonance, 187–188
Marcus relation, 197–198, 204–206
Maximum rate limited by diffusion, 169
Mean reaction time, 14
Mechanisms:
  alternative, 93–95, 99–101, 105–106, 176–
    177
  chain, 135
  for exchange reactions, 52–53
  Lindemann, 160–162
  Michaelis-Menten, 81–82
  parallel-path, 121–122
  reaction, 90–94, 193–206
  S_N1, 73–75
Method(s):
  Bunnett, 58–59
    ad eosdem, 59
    ad eundem, 58–59
  competition, 58–59
  concentration-jump, 48–50
  differential temperature, 132
  flow, 183–184
  fractional-time, 34
  Guggenheim, 25–28
  instrumental, 22–25
  Kezdy-Swinbourne, 25–28
  numerical, 83–84
  perturbation, 184–187
  relaxation, 185–188
  Runge-Kutta, 83–84
  stopped-flow, 183–184
  temperature-jump, 183, 185–187
  time-lag, 24–29
  Wilkinson, 35
Michaelis constant, 81
Michaelis-Menten mechanism, 81–82
Microscopic reversibility:
  forward and reverse rate and, 124
  principle of, 128–131

Net activation process, 175
Net reactions:
  for chain mechanisms, 135
  and reaction rate, 2–3
  and reactions following the rate-limiting
    step, 93
  nmr (*see* Magnetic resonance)
Numerical methods, 83–84

Opposing reactions (*see* Reversible reactions)

Order of a reaction:
  apparent fractional, 5–6
  concentration-dependent, 5–6, 31–32, 91–
    93
  defined, 5
  determined experimentally, 32–35
  fractional-time approach, 34
  interpretation of inverse, 91–92
  log-log plot and, 6, 32
  variable, 5–6, 31–32
  by Wilkinson's method, 35
Oscillating reactions, 143–146

P_∞ (end-point reading):
  in second-order reactions, 23–24
  systematic error in, 24, 68
  unknown, 24–30
Parallel reactions:
  competition method for, 57–59
  first-order, 55–56
  leading to a mixture, 55–56
  mechanistic formulation of, 97–98
  product ratios in, 55–56
  rate constant for one product, 56
  reaction order and, 93–94, 98
Partition function:
  in chain reactions, 141
  expressions for, 154
Perturbation methods:
  for fast reactions, 184–185
  kinetic equations for, 185–187
  temperature-jump, 185
pH-rate profile:
  of consecutive reactions, 105–106
  in preequilibria, 103–104
Photochemistry:
  flash photolysis, 188–189
  in initiation of chain reactions, 134
Photolysis, flash, 188–189
Plots:
  Arrhenius, 119, 121–122, 125–128
  Hanes, 88
  Lineweaver-Burk, 88
  log-log, 6, 32
Predominant species:
  from preequilibria, 102
  in rate laws, 91
Preequilibria, 101–106
  alternative formulations of, 94–95
  involving ion-pair formation, 104–105
  and predominant species, 102
  prior rate-limiting step, 91–92

Preexponential factor, 117
   and $\Delta S^{\ddagger}$, 120, 125
Pressure effect:
   and $\Delta V^{\ddagger}$, 177–179
   on reactions in solution, 177–179
Principle:
   of constant ionic change, 174–175
   of microscopic reversibility, 128–131
Propagation (*see* Chain reactions)
Pseudo-first-order kinetics, 12–15
   (*See also* Flooding, technique of)
Pulse radiolysis, 189–190

Rate laws, integrated:
   for concurrent reactions of mixtures, 57
   multiterm, 35–37
   for reversible reactions, 46
   table of, 30
Rate-limiting step:
   alternative, 96
   in chain reactions, 142
   in consecutive-step reactions, 76–80
   formulation of, 91–93
Rate-pH profile, 103–106
Reaction constant in Hammett equation,
   195–196
Reaction intermediate:
   competition method for, 57–59
   in consecutive step reactions, 65–80
   generation of by flash photolysis, 188–190
   scavenging of, 57–59
   steady-state approximation, 72–80
   versus activated complex, 90
Reaction mechanism:
   extrakinetic probes of, 193–206
   from rate equation, 90–94
Reaction order (*see* Order of a reaction)
Reaction rate:
   defined, 2–4
   and magnetic resonance, 187–188
   (*See also* Initial rates)
Reactions:
   bimolecular, 155–157
   biphasic, 65–72
   branching chain, 142–143
   chain (*see* Chain reactions)
   concurrent, 51–59
   consecutive, 65–72, 92–93
   consecutive-step, 76–80
   diffusion-controlled, 96–97, 168–169
   elementary, 5
   exchange, 51–55, 129–131, 184

Reactions (*Cont.*):
   fast, 182–183
   first-order, 65–72
   heterogeneous, 82–83
   net (*see* Net reaction)
   oscillating, 143–146
   parallel, 55–59, 93–94, 97–98
   pseudo-first-order, 12–15
   rapid, 183–184
   reverse, 123–124
   reversible, 42–50
   second-order, 172–173
   in solution, 166–167, 170
      compared to gas-phase, 166–167
      encounters during, 167
   unimolecular, 159–162
Relations:
   Arrhenius, 117
   Brønsted-Pedersen, 194
   LFER, 194–198, 204–206
   Marcus, 197–198, 204–206
Relaxation methods:
   kinetic equations for, 185–187
   magnetic resonance, 187–188
   temperature-jump, 185
Reverse reaction:
   and microscopic reversibility, 124
   relation to forward, 123–124
Reversible reactions, 42–50
   concentration-jump method for, 48–50
   first-order, 42–45
   higher-order, 45–50
   P_{∞} values and, 44–45
   and rate law terms, 91–92
   rate versus rate constant, 43
   very rapid, 183–187
RRKM model for unimolecular reactions,
   162
Runge-Kutta method, 83–84

S_N1 mechanism, 73–75
Salt effects:
   and ionic charge types, 173
   and reaction mechanism, 176–177
   on second-order reactions, 172–173
Second-order kinetics, 16–21
   equivalent concentrations, 20, 21
   estimation of P_{∞}, 24
   for $k[A]^2$, 16–17
   for $k[A][B]$, 17–20
   when P_{∞} is unknown, 27–29
   using instrumental methods, 22–24

Solvent effects, 170–171

Standard state:
  in solution reactions, 170
  and the value of $\Delta S^\ddagger$, 120, 179–180

Stationary state (*see* Steady-state approximation)

Steady-state approximation, 72–80
  in chain reactions, 142
  mathematical requirement for, 72–73
  simplified derivations, 74, 76–77

Stereochemistry and reaction mechanism, 202–204

Steric factor, 117
  and activated complex theory, 155–157
  and collision theory, 152

Stoichiometric factors, 3–5

Stopped-flow method, 183–184

Successive steps, 98–101
  denominator terms and, 98
  (*See also* Consecutive reactions)

Surface catalysis, 82–83

Taft equation, 194

Technique:
  of flooding (*see* Flooding, technique of)
  quenched-flow, 184
  stopped-flow, 183–184

Temperature:
  coalescence, 187–188
  effect on rate: activated complex theory expression for, 117–120
  activation parameters for, 116–120
  algebraic expressions for, 116–118
  Arrhenius relation for, 117
  collision theory expression for, 117–119
  control, 9, 10

Temperature (*Cont.*):
  magnitude, 9, 10

Temperature jump method:
  apparatus for, 185
  kinetic data from, 185–187
  time resolution of, 183

Theory:
  activated complex, 117–120, 153–159, 170
  collision, 117, 150–152
  RRKM, 162

Time-lag methods:
  for first-order kinetics, 24–28
  Guggenheim method, 25–28
  Kezdy-Swinbourne method, 25–28
  when P_∞ is unknown, 24–29
  for second-order kinetics, 28–29

Transient intermediates (*see* Reaction intermediates)

Transmission coefficient, 117

Turnover number, 82

Unimolecular reactions, 159–162
  Lindemann mechanism for, 160
  in the presence of nonreactive gases, 161–162
  pressure dependence of, 159–160
  RRKM theory for, 162

Volume of activation ($\Delta V^\ddagger$):
  interpretation of, 178
  from pressure dependence, 177–179

Wilkinson's method, 35